T0305056

THE AGE OF LOW TECH

Towards a Technologically Sustainable Civilization

Philippe Bihouix

Translated by Chris McMahon

Originally published in French
by Editions du Seuil in 2014 as

*L'Âge des low tech:
vers une civilisation techniquement soutenable*

BRISTOL
UNIVERSITY
PRESS

Originally published in French by Editions du Seuil in 2014.
English language edition published in Great Britain in 2020 by

Bristol University Press
University of Bristol
1-9 Old Park Hill
Bristol
BS2 8BB
UK
t: +44 (0)117 954 5940
e: bup-info@bristol.ac.uk

Details of international sales and distribution partners are available at bristoluniversitypress.co.uk

© Editions du Seuil, 2014.
English translation © 2020 by Chris McMahon.

British Library Cataloguing in Publication Data
A catalogue record for this book is available from the British Library

ISBN 978-1-5292-1326-3 hardcover
ISBN 978-1-5292-1327-0 paperback
ISBN 978-1-5292-1328-7 ePub
ISBN 978-1-5292-1329-4 ePdf

Cover design: Liam Roberts
Front cover image: iStock / Rouzes

Bristol University Press uses environmentally responsible print partners.

Printed in Great Britain by CMP, Poole

Contents

List of Figures and Tables

Figures

Tables

Acknowledgements

Philippe Bihouix expresses his thanks to Christophe Bonneuil, Hugo Carton, Yves Cochet, Alain Gras, Jacques Grinevald, Sophie Jeantet, Christophe Laurens, Pietro Majno, Laurent Pré, Maxime Oberlé, Agnès Sinaï, Xavier Verne, for their enthusiastic support, their careful reading of the original French manuscript, their suggestions and advice, and for good conversations.

He extends a special thank you to Christophe, for having had confidence in the project, and to Xavier, without whom the section on new technologies would have been much less detailed and thought-provoking.

He also thanks Chris McMahon, for the great pleasure he had in working on this English version, a version that would not have existed without Chris's commitment and efforts.

Chris McMahon thanks Philippe for his enthusiastic engagement in the project and for the great interest he showed throughout the work, which has been a real pleasure for him also.

Chris also thanks Karen Bell, Laura Dickinson, Andy Frayne, Rachel Freeman, Phil Harris, Anja Maier, Douglas McMahon, James Norman and Ian Roderick for their feedback on the English translation, and especially thanks John Brenton for his enthusiasm for the work and for his introduction to Bristol University Press.

Finally, Chris and Philippe thank Leonie Drake, Kathryn King, Paul Stevens and Freya Trand and their colleagues at Bristol University Press for all their help in bringing this edition to fruition.

Preface to the English Edition

A repeated theme in the story of humanity is its relationship with resources. Historically, the human population of the Earth survived on what could be gathered, hunted, fished, then grown on the land, and on energy from the burning of biomass, human and animal labour, and then wind and movement of water. Mineral use was generally constrained by the energy and tools available and by the need to use wood for smelting. This limited the population that could be sustained and the production and consumption per capita.

With the coming of the Industrial Revolution, the use of coal and then, increasingly, oil and natural gas allowed more energy to be expended, more land to be brought into food production, more population, ever-larger conurbations and transportation systems. The increase in industrial production and consumption resulted from an increase in the use of all manner of materials, whose extraction and production were enabled by the use of large quantities of fossil fuel energy.

However, the production of raw materials, their transformation into artefacts, and the disposal of the products at the end of their use led to increasing environmental impacts. It has been clear for some time that a number of constraints impinge upon our ability to maintain such material flows of the 'linear economy' – the availability of material resources, especially fossil fuels, the ability of the biosphere to cope with quantities of waste, various pollutions and emissions, especially greenhouse gases resulting from the combustion of fuels and the impact of agriculture. These constraints threaten the continuation of economic activities and ultimately the ability of the Earth to support the human population.

In order to address the challenges that energy and resource constraints impose, a number of technological solutions have been proposed, including transitions to renewable energy and biotechnologies, the use of 'smart' digital tools to improve the efficiency of resource use, and adoption of a 'circular economy' in which much more extensive use is made of repair and refurbishment and material recycling. At the moment, these different technological solutions are heavily promoted, in particular as a way of addressing climate change, with the view that through research and development and through continuous innovation we can tackle this existential threat while maintaining 'green growth'.

This book was written by Philippe Bihouix to signal his scepticism with such 'high-tech' approaches and to propose an alternative way forward. Following his earlier study of the future availability of metals in an energy-constrained world, he challenged the view that increasingly innovative technical means would have a positive impact on climate, environment and sustainability: because so-called 'green' technologies, such as electronics or renewable energy devices, demand increasing use of non-renewable resources that are often more scarce, and because a true circular economy is far from achievable due to recycling losses and dispersion.

In his view, the concept of 'green growth' also seems both dangerous and absurd over the long term. Maintaining a growth rate of 2 per cent on a global scale implies doubling GDP (gross domestic product) every 37 years – or multiplying it about 390 million times every 1,000 years. Economists are banking on being able to decouple economic performance from resource use and emissions, which involves increasing GDP while simultaneously reducing polluting emissions, waste and resource consumption. That can undoubtedly work partially or for a certain period of time, but it's impossible to believe that, a thousand years from now, we will have been able to create technologies that are 390 million times more efficient or with that much less impact than now. If we're going to capture the efficiency gains from technology, we first have to question our needs and work towards an approach of frugality or sufficiency and a cleverer use of technology.

This argument is developed in this book in four main parts. After an introductory Prologue, Part I begins with a review of how, since antiquity, technology has always responded to shortages of resources, describing the origins of industrial chemistry, of energy technology, of food production and storage, and of construction materials, but then explaining why high-tech solutions are not the answer this time. In Part II, principles of a low-tech approach are developed, founded on questioning needs and rooted in the search for simplicity and conviviality, localization and design and manufacture for true sustainability. Part III then explores what daily life might be like in a low-tech era, considering topics as diverse as agriculture and food, transport, construction, manufactured products, finance, information technologies and love and leisure. Part IV asks the question 'Is transition possible?', exploring some political, cultural and moral questions and concluding that, if transition is necessary, it is certainly possible and we have ample technical, financial, social and organizational resources. A concluding Epilogue develops further a call to action.

When *L'Âge des low tech* was published in France in 2014 it joined a number of works around the world warning that the race to more technological societies was fraught with dangers, and calling for more convivial or appropriate approaches. What perhaps distinguished *L'Âge des low tech* was that it was written firmly from an engineering perspective, with the historical context clearly set out, and with concrete proposals for a future 'low-tech' world ranging from changes to the design of the artefacts we use every day to wide-ranging social and political changes. It was the strong engineering perspective of the work that attracted Chris McMahon to the book when, in 2017, he first came across it and Philippe's writings and interviews. This English edition arose from Chris approaching Philippe to ask if a translation existed or was planned, and subsequently volunteering to produce one.

Six years after the first publication, the issue of climate change is ever more urgent, and receives increasing public attention with the term 'climate emergency' widely used. There are mounting concerns also with plastic pollution and the effects of agro-chemicals and soil degradation, while the issue of peak oil, and similar peaks in production of other limited resources, has not gone away. Yet the numbers – for CO_2 production, for the consumption of oil, coal and minerals, the production of plastics, the destruction of natural habitats – go ever upwards. It is clear that 'business as usual' tempered with 'green' technologies has not been a solution, yet. The message of *L'Âge des low tech* is even more important and relevant than ever.

The book was originally written for a French audience, and we thought when we began the translation that we might need to change a lot to make it understandable and relevant to an English-speaking audience. The readers of the draft translation disagreed and felt that the examples were easy for an international audience to understand. Where French examples were retained, the message that they gave was easily transferred to a wider context. But we did take the opportunity in the translation to update the data presented, and to broaden examples and the text especially to a wider European or worldwide context where we could.

We may have entered difficult times, in which decisions to cope with environmental challenges while maintaining social stability will probably not be easy. We hope this book will be a modest contribution to the finding of a path towards an industrial system and civilization that is really sustainable, from both a technical and a social point of view.

Philippe Bihouix and Chris McMahon, April 2020

Prologue:
The Mad Dance of the Shrimps

It is May 1940. German armoured columns have broken through the French lines and the terrified population is pouring onto roads that are quickly becoming clogged by refugees. From his spotter plane, sent on a desperate mission over enemy lines to gather information that nobody will use, Antoine de Saint-Exupéry contemplates the debacle:

> Of all these objects the most pitiful were the old motorcars. A horse standing upright in the shafts of a farm cart gives a sensation of solidity. A horse does not call for spare parts. A farm cart can be put into shape with three nails. But all these vestiges of the mechanical age! This assemblage of pistons, valves, magnetos and gear-wheels! How long would it run before it broke down?[1]

Forgive me, my dear Saint-Ex, for using your insightful reflection as an example of low-tech, simple technologies. As an intrepid aviator, you were of course wholly convinced by advanced technologies. But you asked for it, proposing the abandoning of yesterday's elegant cars to return to the horse and cart! Nothing, for me, summarizes better the crucial question facing our industrial society. Exchange electronics for the mechanical – replace pistons and valves by transistors and capacitors – and the insight is as fresh today as it was in 1940. Could our technically complex, globalized, specialized world withstand a catastrophe, whatever it might be – a dearth of easily accessible energy and material reserves, the consequences of pollution – especially climate change – or some new and more acute financial and economic crisis?

This book develops the thesis, radical I know, that instead of seeking top-down solutions to current environmental and societal challenges, instead of seeking ever more innovation, high technology, digitalization, competition, networking, growth – giving them names such as 'sustainable development', 'green growth', 'Economy 2.0' – we must instead direct ourselves, as quickly as possible, to a society based primarily on simpler technologies, undeniably cruder and more basic, maybe a little less powerful, but much more resource-efficient and locally controllable. This idea is not new: as early as the 1940s and 1950s, writers like Bernard Charbonneau and Jacques Ellul[2] denounced the race towards more technological societies; in the 1960s and 1970s, Ivan Illich[3] and Ernst Friedrich Schumacher[4] argued for the use of 'convivial' or 'intermediate' technologies. More recently, authors like Langdon Winner[5] and John Michael Greer[6] developed the same kinds of ideas, while Kris de Decker launched a very comprehensive website dedicated to historical analysis and refreshment of past knowledge and technologies, now also published as a book.[7]

Before going into detail, I owe readers some explanation about what led me to take such a view. Nothing predestined me to choose the horse and cart, or to take the opposite view to the majority of my fellow engineers, who swear by high technology, research and development, and innovation. In short, I should explain why I must take a view opposite to today's conventional wisdom and in contradiction to those who assume unstoppable progress.

Born two years after the first moon landing, my childhood, like many of my generation, was marked by many scientific and technical exploits, entertained by sci-fi films and regularly filled with 'revolutionary' products. In the year that I was 10, Space Shuttle *Columbia* took off from Cape Canaveral – the poster is still on the wall of the room I had as a child – and a few months later *Paris Match* published the superb images of Saturn that were transmitted by the *Voyager 2* probe. At the beginning of the 1980s, the first wave of consumer electronics began, with electronic calculators, the first Japanese digital watches with their tiny lithium batteries, and handheld video games. As school students, we spent hours programming low-resolution arcade games like Space Invaders on early computers provided by the Ministry of Education (provided, I suppose, as support for French technology and for the recently nationalized Thomson-CSF company against its great rival Amstrad – the processor was a Motorola, but hey). And soon the Sony Walkman would let us experience the joys of personal music on the move.

In short, life followed its course, and progress its obviously linear path. There had of course been some technological disappointments. The popular science magazines of the 1950s were a little premature in announcing electricity that was too cheap to meter, nuclear cars and toasters, and even helicopters for urban travel. And, contrary to predictions, supersonic aircraft did not cross the oceans in their hundreds – two oil shocks put a stop to that. But, with the oncoming rush of new information, only a grumpy few remembered.

Of course, not everything was perfect on the planet. Developing countries had not developed as fast as expected, but everyone suspected it was partly their fault anyway. Decolonization was still recent, and 'technology transfer' programmes were in full swing, against the backdrop of the Cold War. The people of the Soviet bloc seemed to be having a bit of a hard time, but that made great scenarios for spy movies. There was pollution, but it was localized, at least in people's perception. Yes, the mercury poisoning of Minamata Bay was horrible, but it didn't affect so many people and it was a long way away. One might even say that localized pollution, pollution 'in our backyard', was on the wane.

That was indeed sometimes the case, as a new phenomenon had appeared that partly accounted for lower levels of pollution and that was promised a bright future: deindustrialization. We saw the effects across Europe, especially for coal mining and areas with particularly visible industries such as blast furnaces and metallurgical plants. In the case of the iron and steel industry, deindustrialization was largely a rationalization of productive apparatus, and a downward adjustment in capacity that followed from a reduction in demand. The effort of post-war reconstruction, the era of growth from the late 1940s to the 1970s had passed. For coal mines, it was more a matter of the closure of mines that were no longer profitable. But a new trend, deindustrialization by relocation of production, was taking shape imperceptibly. 'Made in Europe' was moving on to other places. From the 1970s, Japanese products began to roll back 'Made in the USA'. European industry also turned to the east as city-states like Hong Kong and Singapore were beginning their success stories, backed by a China that was preparing to become the factory of the world.

Well, you know the rest. While the fall of the Berlin Wall brought hope of a bright future, the global impacts of human activities emerged in public debate: the hole in the ozone layer, deforestation everywhere continuing apace, then, and soon enough, climate change. This time, things were really starting to look not so good for the planet. For a while, at the beginning of the 2000s, such matters were eclipsed by the first internet madness and the 'dematerialization of the economy', but questions soon returned. Recall French President Jacques Chirac saying in South Africa in 2002: "Our house is burning and we are watching elsewhere. [...] We will not be able to say that we did not know."[8]

While all this was going on, I was getting my education. I had learned at school how to solve equations a little faster than some others and had become a typical product of French meritocracy, promised a bright future in science and engineering, even if my engineering Grande École (college) had begun, like many others, to graduate battalions of traders, financiers, auditors and consultants. A few years of industrial experience, though, allowed me to discover the material reality of our economic system and its physical consequences. 'Environmental' approaches could have only limited effect and European and global integration was under way. The 'mad dance of the shrimps', in which shellfish were caught in the North Sea and then, for reasons of labour cost, shipped to Morocco to be shelled, or strawberry yoghurts for which, in 1992, the ingredients had travelled more than 9,000 km, helped engender in me a certain scepticism concerning the nature of progress.

Fortunately, of course, the concept of 'sustainable development' was to arrive to save the day. Amid a surge of activities and publicity the concept was formalized in the Brundtland Report in 1987 and pioneered the response to planetary challenges. Like me, you will have noticed that everything now has become 'sustainable'. There is no product that is not 'eco-designed', no city development that is not an 'eco-neighbourhood', no building of any importance that is not 'low consumption' or 'environmentally friendly'. Even roads, airports and Formula One races are now being declared environmentally friendly, thanks to measures to protect toads from being squashed, or because of progress towards more efficient engines. All major companies and local communities produce thick reports – which were originally on glossy paper and are now from 'sustainable sources' – to present their strategy 'for sustainable growth', to promote their commitment to the planet and to present their key data which are of course all 'green'. This is the time of the 'circular economy™' and 'industrial ecology©', astounding oxymorons and idols for modern times!

Well, we have gone at it for a few years now with little effect. We have cut down trees and burned oil in order to explain to ourselves that we were going to protect forests and economize on fuel. The discourse of sustainable development has been over-used, twisted, diverted, degraded, become ridiculous enough to make us sick. But facts are stubborn, and, like any engineer, I like facts and figures. In reality we have never produced, consumed and discarded as much as we do now. Bees take refuge in cities, preferring diesel soot to the 'innovative' molecules of the agro-chemical – sorry 'phytosanitary' – industry. Our rubbish bins are full and overflowing. Even if the weight of the rubbish can be reduced a little, its harmfulness increases – and recycling rates are progressing at a sluggish pace.

Many European countries consider themselves virtuous, in real environmental transition. But consider what is happening in France (see Figure 0.1; with variations, we will see similar figures for many European countries). Its people produce around 2 tonnes of industrial waste per inhabitant per year, almost 5 kg per person per day! Each day, its

Figure 0.1: One day in France ...

Apparent daily consumption and waste per person

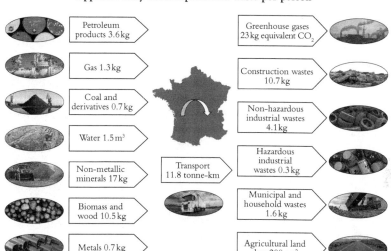

Note: Balance of imports/exports based on semi-finished and finished products.

Sources: Institut Francais de l'Environnement (French Environment Institute – IFEN)
French Government Ministries, Agence de l'environnement et de la maîtrise de l'énergie
(Environment and Energy Management Agency – ADEME)

inhabitants generate 12 tonne-km of freight movement per person – that is about 100 kg of goods moved an average of 120 km – 88 per cent by road. As the urban sprawl shows, about 1 per cent of the country – the size of a French administrative *département*[9] – was built on in ten years, and then a further 1 per cent in only seven years! This has often been on the best farmland, resulting in an irreversible mess – nothing edible will be produced for hundreds of years under the tarmac of supermarket car parks.

On a global scale, 20 per cent of the world's population continue to exploit more than 80 per cent of its resources, and in one generation we are about to extract more metals from the Earth's crust than in the entire history of humanity. It is tempting to blame emerging economies, and China above all, for this increase. But let's not forget that China's resource consumption is also driven by its role as factory of the world, and we import, directly or indirectly, a good part of its output.

The world is like a movie set. The façades, for the consumer, still attempt to look good. In advertisements, in shops on supermarket shelves, everything is fine. But behind the façades there is a reality. There are hidden consequences even when we have the best of intentions. I can buy a mobile phone in France, and in doing so I will have exploited Congolese miners, destroyed primary forests in Papua New Guinea, enriched Russian oligarchs, polluted Chinese water tables and then, in 12 to 18 months' time, I will dump my electronic waste in Ghana or somewhere similar.

The world is like a kind of huge espresso machine, like those that are so emblematic of our economic and industrial system, and in which capsules of used coffee-grounds disappear into the bowels of the machine. The waste is stored there until the bin can be swiftly and discreetly emptied out of sight. For those who can afford it, even that task may be done for them by a cleaner. And while this goes on, poachers pursue the last elephants, the remaining primary forests are transformed into paper tissues (Tasmania and Canada), into plywood and oil palm plantations (Indonesia and Malaysia) and into transgenic soybean plantations (Brazil and Argentina), oceans are becoming covered with plastic debris and land and water are becoming permanently poisoned by pesticides. There is little to be proud of.

Faced with probable constraints on resources in the future – renewable resources (such as fish stocks) as well as non-renewable resources (fossil fuels and metals) – we behave like Molière's doctors. These were the kind of physicians the English came to call Leeches. They believed that bleeding was a cure, and if the patient got worse, then not enough blood had been drawn – until the patient died, needless to say! For us, we swear by innovation and technology. Millennia of exploration, experiment and innovation brought us to the incredible acceleration of the 19th and 20th centuries. This came at the cost of pollution and has now brought us to unprecedented social and environmental destruction, the ultimate consequences of which we can no longer know. We even have trouble admitting to ourselves the consequences in the here and now. Faced with the unknown, with the troubling, brutal disease of the Anthropocene, we still hope, or pretend to hope, for future growth and that this can become 'green growth', to reverse the effects of our past activities. We even seem to say, absurdly, that such growth should accelerate because 'a little growth pollutes, a lot of growth cleans it up' (thanks to the boost of innovation).

I do not believe it. Not any more, for the many reasons I have shared with you in this Prologue. But rest assured, I don't propose that, as in Dante's *Divine Comedy*, you should abandon all hope (*Lasciate ogne speranza, voi ch'intrate*). On the contrary, I believe there is a way of avoiding a global crisis, conflict and collapse, or, more simply, depression and despair. And who knows, maybe an age of low tech, a time of a technologically sustainable civilization, will come.

PART I

The Rise and Fall of
'Engineering Miracle-Workers'

Innovation, we are told: research and development, green growth, high-tech and of course clean and resource-efficient products: these are the answers to all ills of the planet and of our societies as we search for a new lease of life. Sleep well, good people, because in our leading companies and ultra-modern laboratories painstaking researchers and engineers in white coats work and invent for you. We research, and, because we research, we will find, because we always have, haven't we?

Are we not finding oil and gas ever deeper under the oceans, not to mention by fracking which of course will be exploited in an ecological way? Breakthroughs in 'clean' cars and 'green' technologies are imminent. According to the most optimistic among us, we are at the dawn of a third or fourth industrial revolution. Thanks to *smart grids*, and *intelligent* energy and transportation networks (built on a model similar to the internet), we will all become producers and storage providers of green electricity. We will move to a hydrogen economy, to a circular economy that will recycle its wastes thus making available new resources; a future based on ever more mobility and connectivity ...

Thus, at the risk of provoking a certain cognitive dissonance[1] in the attentive reader, listener or viewer, we find that dire observations in the serious media concerning the state of the planet alternate with grand announcements of new technological breakthroughs, inventions, pilot projects and amazing start-up companies. We hear of freight-carrying airships and solar aircraft, of offshore wind turbines and solar plants in the desert, of wave and tidal power. Solar buildings will produce more energy than they use, 'air purifying' paints will clean up your indoor air. And that is not to mention the pages of discussions about 'green consumption' or the 'sustainable economy' that hardly consider the real environmental impact of the products on view: for example a mobile phone in a bamboo

1

case or a low-consumption headlamp with integrated dynamo specially for 'eco trekking'.

As the planetary indicators show, the present reality, the truth[2] is that the world has never been so polluted and pillaged, even if, compared to the 19th century, there have been local improvements regarding some pollutants and industrial wastes. And yet, paradoxically, according to journalists, economists and scientists, we have never been so chock-full of the technical means for environmental management, never possessed so many remarkable 'ecological' inventions.

But, despite all of the announcements, the truth is they struggle to deliver. It is as if our societies, faced with an ominous situation, have a need to take refuge in almost messianic attitudes, in the promise of paradise the day after tomorrow if we survive the present torment. The key to this seems to be a discourse that cannot be aligned with reality and which has led, at least for the moment, to a world of permanent contradictions, in Bertrand Méheust's words to a *politics of the oxymoron.*[3]

What is the reality behind the façade? Certainly, it is true that scientists, engineers and engineering businesses – our illustrious predecessors – found solutions, often seemingly almost miraculous, to the challenges of resource shortages in their day. Provided, as always, that one does not look closely at the natural destruction or pollution to which these led.

How technology has (always) responded to the shortage of resources

The history of humanity is that of a long struggle with resource scarcity. Every species is constrained by the availability of resources in its environment. It is a principle of Darwinism that natural selection in the face of constraints is, with mutation, the engine of evolution. But humans are almost the only creatures to use exosomatic tools, to extend beyond the limits and constraints of their bodies. Instead of using claws, teeth or bristles, the human uses sharp tools to hunt, contrives clothing for protection against the cold, and uses heat to aid in the digestion of food.

Of course, if food was sufficient, as long as these tools remained rudimentary, and population numbers were limited, then scarcity of tool-making resources would seem to have been a rare constraint. Oh, lucky Palaeolithic man, with immense herds of reindeer or horses – and maybe an occasional mammoth or woolly rhinoceros as a treat – to provide for all his needs: fat for light in the dark, bone for small tools and decorations, skins, furs and tendons for clothing. Maybe a little ochre picked up from the ground for make-up, and a fire-hardened stick or a sharpened stone as a weapon. So, was it an age of abundance?[4] We cannot be so sure. Even back then, access to some tool resources could be critical. We know that some flints travelled several hundred kilometres, and that some mines supplied more remote areas. At that time, exploration was required to meet certain basic needs: already some objects were available only in limited quantities.

As for food resources, after perhaps moments of euphoria, there were generations who had to live through times that were particularly cruel or demanding of ingenuity, at least from a purely Darwinian point of view. The use of fire for hunting quickly deforested huge areas of dry forest and savanna, harming many animal species. Humanity is most likely to be responsible for the extinction of much of the world's megafauna, especially in Australia (around ~60,000 years ago) and in the Americas (around ~11,000–14,000 years ago), rather more so than in Europe and Africa where animals co-evolved with hunters and learned to be wary of them. On islands, events are more recent and the responsibility of humans is clearly established: the dinornis of Madagascar and moas (giant ostriches) from New Zealand, dwarf hippopotami from Corsica or Cyprus and so on.[5] The Palaeolithic era probably had periods of orgiastic feasting – think of tribes of hunters after long journeys coming across hordes of animals that had not yet learned to be afraid of them – interspersed by lean years.

It is impossible to precisely date the first shortages of non-food raw materials. Concerns must have begun to arise as soon as men began to use materials that were less abundant than stone, wood or clay, that is to say around ~6,000 years ago, with the first metal tools made of copper and, to a smaller extent, iron from meteorites, a natural ferronickel alloy with good corrosion resistance. Early developments in metallurgy, from copper to bronze (an alloy of copper and tin) then made it possible quite rapidly to exploit ores, oxides and carbonates, not just metals in their native state, and thus to meet growing demand.

With the growth and concentration of the human population and the development of great civilizations, as in the Middle East and around the Mediterranean, the first shortages, mainly local and regional, would have presented themselves, and not just in non-renewable resources such as metals. There was an over-exploitation of a priori renewable resources (forests, soils, animals): Phoenician sea snails from the Muricidae family, used to produce the dye used for the imperial purple band on the togas of Roman senators (and the entire toga for triumphant generals and emperors), have almost disappeared from the Eastern Mediterranean today, and the corresponding dye production technique has been lost.

Migrate, trade or invent

There are three strategies that may be used, alone or in combination, to cope with shortages. One may *move location* – temporarily through nomadism or permanently through migration – which is an effective method to avoid local shortages. One may exchange local surpluses for other goods through *trade*, which helps rebalance inequalities of resources in different places. Finally, one may *invent* – find a way to produce the missing resources from another source, or learn to do without it – usually by finding a substitute.

All three strategies are still used today. When some starry-eyed dreamer proposes that we mine metals on the moon or use the CHON (carbon/ hydrogen/oxygen/nitrogen) comets of the solar system as a source of food, they only offer us a new type of nomadism, on a different scale. To remain believable, they are not proposing, yet, that we trade with little green men. But while we wait for extraterrestrial solutions, because migration and commercial strategies have de facto been limited by the exploration and colonization of the entire planet, we are necessarily trusting our future to technological innovation.

To what extent have past shortages and appetites been a spur to the exploration of the planet, to the development of trade and commerce and to the discovery of new techniques, and how does this relate to anthropological fundamentals – to curiosity, the thirst for knowledge and understanding, the desire for adventure, enrichment and glory? Regardless, we can paint a picture of a thousand-year long 'struggle' of technology against the exhaustion of resources, whether they serve us for heating, clothing, habitation, food, transportation or amusement. This struggle has accelerated with the pace of innovation in the past two or three centuries, with a consequent spectacular growth in the consumption of resources.

Energy is paramount

Let us recognize what is of central importance. The availability of energy is paramount, since it is usually necessary for the transformation of other resources (smelting or heat treatment of metals, transformation of materials such as clay by firing in a kiln, activation of chemical reactions by heat and so on), and for their transportation to the place of consumption.

The overall history of energy is quite well known. Until the end of the 18th century, wood was by far the most important source of energy, supplemented by wind and water mills and animal traction. Contrary to popular belief, these four sources remained very important until late in the 19th century, even in the most industrialized countries, before giving way to fossil fuels.[6] The exploitation of forests for fuel and timber left deep (and in many cases lasting) marks on European landscapes, and from the 17th century a wood crisis was becoming widespread. Coal had been known and used sporadically for centuries, especially in England where London's air pollution was legendary. The double invention at the turn of the 18th century of the steam pump (by Thomas Savery) and the steam engine (by Thomas Newcomen), then their improvement by Denis Papin, James Watt and others, allowed water to be pumped out of mines and thus greater resources below groundwater level to be accessed.

From there we know how it went. The great adventure of the Industrial Revolution began: factory production, mining towns and villages, Great Britain exporting its coal to the world, then came the time of oil and gas, finally hydroelectricity and nuclear power. This is of course a great simplification, owing to our linear representation of technical progress, because we have never been out of the coal age. Since the first tonne was extracted, production and world consumption have increased continually, even through economic crises. We are producing about 7.7 billion tonnes per year (coal and lignite) in 2018, which makes coal the second largest source of energy consumed (3.8 billion tonnes of oil equivalent, or Btoe), just below oil (4.7 Btoe) and above natural gas (3.3 Btoe) (see Figure 1.1). Among the countries with the largest consumption we find, besides China, are high-tech countries like the USA and Germany.

Oil was not the solution to a shortage of coal, but of whales! At the end of the 19th century (oil exploration began in Pennsylvania in 1859), much domestic but also public lighting still used whale oil. Moby Dick and his friends then breathed their last because technical innovation and the enthusiasm of whaling captains played havoc with their numbers: steam propulsion and the harpoon cannon led to the virtual extinction of right whales and sperm whales. Humpback whales were still numerous and were, until then, inaccessible because they were faster and especially because they sank after death (but they would soon meet the same fate, in the early years of the 20th century, thanks to other innovations in whale hunting).[7] Petroleum was therefore first used as a lighting oil and lubricant, and production was artisanal – 30,000 'family-owned' wells requiring little in the way of equipment and investment, each producing about a barrel of oil a day. But then, in the winter of 1901, a 'gusher' at Spindletop in Texas suddenly tripled oil production in the United States. New applications were needed quickly. Oil was used for the generation of electricity, but it was especially the spectacular growth in the use of the internal combustion engine – in particular from 1908 with the launching

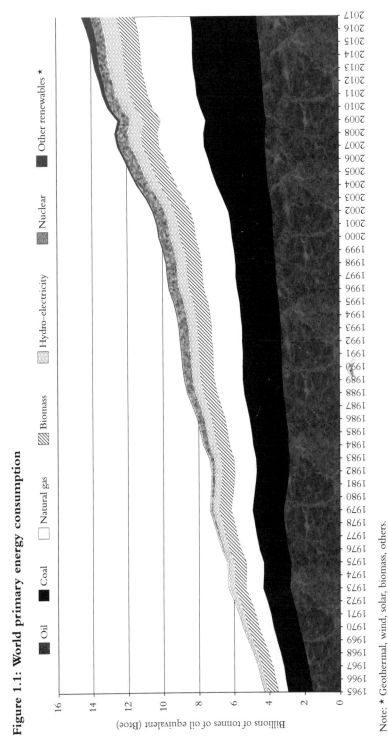

Figure 1.1: World primary energy consumption

Oil   Coal   Natural gas   Biomass   Hydro-electricity   Nuclear   Other renewables ★

Billions of tonnes of oil equivalent (Btoe)

Note: ★ Geothermal, wind, solar, biomass, others.
Source: BP Statistical Review of Energy, 2018

of the famous Model T Ford – which made it possible to find an outlet for this abundant and cheap oil.

As for nuclear power, with less than 5 per cent of the world's primary energy production, it remains a secondary source (but not, unfortunately, in terms of the potential risks and damage that it might cause) that has only been developed in close connection with the military-industrial complex. Civilian power plants were needed to produce the plutonium needed for the many H-bombs from their uranium fuel.

Using metals like a swarm of locusts

In the case of metals, it is mainly the 'locust' mining strategy of stripping everything bare and moving on that has made it possible to cope with ever-increasing demands. Nowadays, dozens of mines are abandoned each year while production has to be started at many others to maintain or increase the quantity of ore extracted. Some countries exhausted their ore deposits quite early, sometimes even in ancient times, while in other cases, although there is still ore in place, producers have turned elsewhere to higher quality reserves that may be worked at lower cost.

The example of precious metals illustrates this phenomenon quite well. Athens was able to arm its fleet against Persian invaders thanks to the legendary mines of Laurion, which produced silver, copper and lead, but which were largely exhausted from the 3rd century BCE. Gaul, and especially Spain then Dacia (in present-day Romania) were great sources of gold for the Romans. Control of the Spanish mines was absolutely strategic for Rome, with its great need for gold and silver to fund a trade deficit with the Orient arising from the ostentatious habits of its richest noblemen and military leaders. Spain, together with Sicily (a granary at the time), was at the heart of the conflict in the Punic wars. *Cartago delenda est* – Carthage must be destroyed – to allow the continued import of silk from China, precious stones from India or myrrh from Arabia.

After the exhaustion of European mines in the first centuries AD, gold became very scarce: the only source was in West Africa, the former 'Sudan' or western Sahel, a vast area ranging from Ghana – formerly aptly named the Gold Coast – to Mali, where it was traded for salt by Arab traders and brought back by caravans crossing the Sahara. Most European countries minted their coins in the more abundant silver, especially after the discovery of the Erzgebirge (literally 'ore mountains') mines in Germany and Bohemia. The conquistadors were therefore not looking for gold by accident. They did find a little gold in the treasures accumulated over centuries by the local *caciques* (leaders), but especially they found silver in the fabulously rich mines of Potosi in Bolivia and Zacatecas in Mexico. There was a series of discoveries and gold rushes in the 19th century, in California with the forty-niners (1849), Australia (1851) and Jack London's Klondike (1896), not to mention Brazil, Siberia and South Africa. Today, while some of these countries are still important (Australia, Russia, South Africa, the United States ...), the gold frontier continues to move: China, Indonesia, Papua New Guinea ...

We have the same experience for almost all other metals: who remembers Cornish tin, the supply of which Julius Caesar wanted to secure (bronze is an alloy of copper and tin, but the latter is much less common)? It was dethroned by the reserves of the east Asian tin belt of Malaysia and Indonesia. How about the 'minette' source of iron from Lorraine in France? Bulk carriers now land their cargoes of iron ore from key players like Brazil or Australia. Or copper from Cyprus, the island of Aphrodite, now better known for its banks and barbed wire than for giving its name to 'the metal of Cyprus' (aes cyprus, later shortened to Cuprum)?

However, in the case of metals, technological breakthroughs have also made it possible to considerably assist exploitation, to recover less concentrated or less accessible resources, and therefore to increase reserves. First, explosives made it possible to blast rocks more easily. Black powder was first used in the Freiberg area about 1650, then nitro-glycerine and dynamite at the end of the 19th century. These were used in place of wood fires and chisels. Compasses permitted determination of the direction of galleries and optimization of the amount of waste rock (overburden material) extracted. Then, the steam engine was used to pump water, as in coal mines. Finally, the diesel engine and mechanization, when possible, increased the energy expended per kilogram extracted, but made it possible to remove much more burden and eventually to use open-cast mining.

Finally, *substitution* has a role to play. In the 18th century, cobalt was obtained from the mines of Thuringia (Saxony) for the production of 'cobalt blue' dye – an alternative to expensive ultramarine, made from lapis lazuli powder from Afghanistan. Cobalt was found in silver and copper mines and, although its ore looked similar to the ores of those metals, it was originally of much less commercial value. It gets its name from *kobald* – the name, together with *nickel*, of underground goblins that 16th-century German miners thought tried to trick them into mining worthless ore. The development of artificial ultramarine in the 1930s drastically reduced demand, while it was nickel that found industrial application in plating and stainless steels, but only at the end of the 19th century with the discovery of rich deposits in New Caledonia. Its by-product cobalt, meanwhile, had to wait for the development of aeronautics and superalloys, then rechargeable batteries, to again attract significant industrial interest.

However, there is virtually no metal whose consumption has declined in recent decades, with the notable (but temporary) exception of lead, with the discontinuation of tetraethyl lead as an anti-knock agent in gasoline engines and the particularly effective recycling of lead-acid batteries, and perhaps mercury, which is and should be progressively replaced in its various uses for reasons of its toxicity. Metals have been replaced in many applications, for example by plastics or composite materials to make structures lighter, but without stopping an explosion in global demand, which has more than doubled in the last 20 years and continues to grow, despite the financial crisis.

From renewable resources to industrial chemistry

Until the late 18th to early 19th century, metals were virtually the only non-renewable resources used (with the exception of a bit of coal, as we have seen, or bitumen from Mesopotamia to seal the seams of boats and to use in Greek Fire incendiaries), besides mineral products available in almost unlimited quantities such as limestone or clay, to which we will return when we discuss building materials.

Animals and plants

For the most part, industrial and consumer products were based on many animal and plant products: natural dyes (madder, pastel, indigo, henna, weld, lichen ...), fats, glues and tallow candles (made from animal and bone waste), alkaline products (soda and potash), acids and alcohols produced by fermentation (acetic acid from vinegar), hides and furs, fibres (wool, linen, cotton, hemp) ...

Limits to capacity were purely related to available land, whether the crop was cultivated or wild. Limits related to the productivity of the land and its carrying capacity in arable or livestock farming, and to the competition between uses for cultivated areas, pastures and woodlands. Linen was needed for sailcloth, hemp for ships' ropes, pitch – produced by distillation from softwood – to seal their hulls. A not-inconsiderable 'farming and forestry hinterland' was needed to be a maritime power. The problem had been experienced since antiquity: early Greek cities had to stock up distant lands with wood and various 'industrial' products when settling colonies in, for them, relatively untouched areas such as on the Black Sea coast. In France, under the *ancien régime*, before the French

Revolution, the colour of your clothes made it possible to determine your social origin at a glance. To avoid trouble, it was better not to pick a fight with any character dressed in the garish and prominent colours that would indicate the purchasing power – and therefore probably power full stop – of the owner.

For Western Europe too, commerce and constantly expanding frontiers made it possible for supply to keep up with increasing demand. Cochineal, indigo and leather were the three main export products, apart from silver, from Spanish America in the 17th century. Buccaneers – mistakenly confused with filibusters or freebooters – hunted the wild cattle that had proliferated on the island of Saint-Domingue, and sold to passing ships the meat, smoked for long hours over wood fires, and also the skins.

Until late in the 19th century, even the beginning of the 20th century, animal resources were at the heart of industrial production. Moulds for jewellery were made with cuttlebone, sperm whale oil was used as a lubricant for machines, cotton gin roller covers were made with the stomach wall of walruses – at the end of the 19th century, 250,000 were slaughtered each year to meet the needs of the cotton industry[8] – not to mention of course whalebone for corsets and umbrellas. From 1900 to 1910, after the discovery of the hydrogenation process, the fatty acid of the whale oil was used to make soap and margarine. By heating, glycerine – the basic ingredient for the manufacture of dynamite – was also produced as a by-product. We can already see the remarkable 'systemic' aspect of our industrial world: the exploitation of whales allowed the supply of essential lubricant to steam engines, and thus, thanks also to the explosives of Alfred Nobel, the extraction of new mineral resources.

Trade in and manufacture using furs was a major source of exploration and Western settlement of the new continents, especially North America and Siberia. In France, in the 1930s, ragpickers, precursors of the recycling industry, still sought from households rabbit skins (from which they could make hats or, more prosaically, glue), feathers and old rubber. Fur was an important export product for countries covered by large boreal forests. But again, over-exploitation and deforestation rapidly reduced populations of martens, squirrels, ermines and other fur animals. The economist John Kenneth Galbraith tells how settlers in America at the beginning of the 17th century, faced with a local cash shortage, adopted for a time *wampum*, the currency of the local Indians, consisting of black and white shells.[9] Its value was guaranteed because it was exchangeable for beaver pelts with the Indians, who thus played a sort of central bank role. Alas, deforestation and hunting caused the beaver population to collapse in one or two generations, and the local currency also saw its value collapse. It was just one of the many recurring financial crises in the United States.

Plant and animal products are still very present as constituents of many industrial products: collagen from animal or fish bones in carpenters' glue, oil from seals or sea lions for the care of shoes, linseed oil, and so on. But these are now limited to specific uses and in modest quantities, of course with the exception of textiles where (in the form of cotton and wool) they still predominate, along with artificial fibres. It is above all the discoveries and developments that followed in industrial chemistry, inorganic and organic, that made it possible to cope with growing demand.

The birth of inorganic chemistry

To put it simply, inorganic chemistry is based on inorganic, mineral products, such as salts, as opposed to organic chemistry based on carbon compounds. It responded primarily to the crucial needs for acids and alkaline products such as soda ash (discussed further later). Acids, especially nitric acid (*aqua fortis*) and sulphuric acid (*vitriol*) were used in many craft and industrial processes: in metal manufacture, in the preparation of dyes, fibres, furs and many more. Alkalis, soda and potash, were used for detergents (in the manufacture of soaps and laundry products), the manufacture of glass, the degreasing of wool and so on.

Acids were originally expensive and produced only in small quantities. Nitric acid was made by distilling saltpetre (potassium nitrate) – a resource intimately linked to living organisms as it is produced by the activity of bacteria in humid cellars and compost. Sulphuric acid was made by collecting sulphur dioxide from burning sulphur or roasting sulphur-containing ores in lead-lined chambers, or by distilling iron sulphate.

'Natural' soda, sodium carbonate, also known as soda ash or washing soda, was obtained from the ashes of certain plants that are rich in sodium, such as salsola soda (hence the name), more commonly known in English as the saltwort, the salicornia (a plant growing in salty soil, especially found in Provence, hence the ancient craft of soap-making in Marseille), or algae. In 1681, Colbert granted the Manufacture Royale de Glaces de Saint-Gobain[10] exclusive rights to the collection of kelp for 25 years along the coasts of Cap de la Hague in Normandy, a first industrial occupation for the region, now more known for nuclear fuel reprocessing facilities. Before the discovery and exploitation of the great mines of Alsace and Germany, potassium carbonate, or potash, was recovered from wood ash, particularly from softwoods, and was imported from forest-rich countries such as Russia and North America when local production was inadequate. With some precautions, chimney ash could be used for washing and cleaning floors, and as a basic fertilizer.

At the end of the 18th century, with the population, and especially the production of glass growing, shortages, dependence on imports and conflicting application demands (farmers used seaweed as a fertilizer and swore at the soda collectors) had made the situation untenable. There was much research to try to produce artificial soda from salt, once the element sodium had been identified, such that in 1781 the French Academy of Sciences even offered a prize for the first demonstration of an economically viable industrial process.

It was Nicolas Leblanc who, at the time of the French Revolution, succeeded in developing the first industrial process for the production of sodium carbonate from salt, sulphuric acid and chalk (calcium carbonate). This process was very polluting and energy intensive.[11] However, combined with the new availability of coal, it opened up an era of abundance for previously rare products. Through its many possible by-products and also some new products, it became the cradle of all inorganic chemistry, before itself being replaced in 1863 by the more efficient and less costly Solvay process.

By allowing the large-scale exploitation of mineral resources for their transformation into industrial products, the Leblanc process changed the scale of industrial pollution: in particular it led to the dumping of large loads of hydrochloric acid in industrial neighbourhoods. Of course, pollution, especially of watercourses, existed before the development of inorganic chemistry. Medieval towns with craft industries tried, with difficulty, to reconcile the use of water for domestic purposes with the unpleasant waste products of tanners and other leather-workers, washer-women (sorry for the assignment of a gender role, but that is who they were), soap-makers and dyers,[12] while the air was often thick with smoke from the burning of wood and coal. But this pollution, although sometimes severe, was essentially limited to highly urbanized areas. Discharges from the first chemical plants also reached the countryside, covering an area proportional to the height of their chimneys.

The explosion of inorganic chemistry

Apart from acids and alkalis, the other major industrial demand was for dyes. As we have seen, these were plant or mineral products, chemical compounds of metals: vermilion and orpiment – respectively sulphides of mercury (cinnabar) and arsenic – white lead, and ochres which are oxides of iron, and so on.

Until the end of the 18th century, dyers were mostly content with plant pigments (mineral pigments were largely reserved for paints and coatings). It must be said that the annual production of cotton fabrics, even in a country at the forefront of production like England, did not yet exceed the equivalent of one shirt per person per year – and most were white. But both demand and production would soon explode.

Organic chemistry, the chemistry of carbon – essentially that embedded in fossil resources like coal, gas and oil – made great theoretical progress in the first decades of the 19th century. In parallel, the industrial distillation of coal was being developed rapidly. It was used to produce coal gas for lighting – another technical solution, used in parallel with whale oil – 'town gas', mainly composed of methane and carbon monoxide, which was replaced in the middle of the 20th century by 'natural' gas from oil and gas fields. Coal gas plants, with their distinctively complex pipework, flourished in many urban areas from the 1820s, and it was important to understand and master the corresponding chemical processes. The distillation of wood and coal was a source of many different organic molecules, which formed or were concentrated in tars. Benzene (a closed chain of six carbon atoms with one hydrogen atom attached to each) and its derivatives such as quinoline were discovered in this way.

The most important derivative was aniline – derived from benzine and naturally present in indigo – which is the basis of organic dyes called azo dyes. From the 1860s numerous chemical industries were founded, especially in Germany, to produce such organic dyes. Most of the big conglomerates of organic chemistry (BASF, Bayer and others) had their origins in the dye industry. The acronym BASF, name of the second largest chemical producer in the world, stands for *Badische Anilin und Soda Fabrik* – Baden aniline and soda factory. In that name we find the two flagship products, the two pillars of organic and inorganic chemistry.

Another revolution took place from the 1910s but especially from the 1930s: polymerization, which would give rise to artificial materials (plastics, synthetic fibres, resins and glues, and so on) that could complement or substitute for leathers, natural textile fibres (flax, hemp, cotton …), animal products, wood and metals. Synthetic polymers have given access to a hitherto unimaginable amount of resources, derived from oil and gas, for the interiors of our buildings and vehicles, everyday consumer products, packaging, clothing and many more.

Bakelite (based on phenol and formaldehyde, patented in 1909) and synthetic rubbers were the first artificial polymers to be used in large quantities. They came at just the right time, together with latex-based natural rubbers, to meet the growing needs of the automobile industry. They were followed by many others, discovered and commercialized starting from the 1930s and through to the 1950s. Today, world polymer production is in the order of 360 million tonnes per year (2018 figures), with a growth of ~5 per cent per year over the past 20 years. Ninety per cent by weight is concentrated in five families of materials: polyethylene (PE: packaging films, bottle caps), polypropylene (PP: bumpers, car dashboards), polyvinyl chloride (PVC: window frames, pipes), polystyrene (PS: yogurt pots) and polyethylene terephthalate (PET: bottles). If we add polyamides (car sunroofs and nylon stockings), polyurethanes, polyesters and butadiene polymers (vehicle tyres) we cover just about everything.

Unfortunately, these materials, for the first time, had a great disadvantage when compared with wood or rabbit skins: they were not biodegradable and therefore generated unprecedented pollution, not to mention their environmental impact at the manufacturing stage. In Europe – a rather 'good' performer compared to the world average – 25 per cent of plastic waste is now disposed of in landfill, 42 per cent is incinerated (sorry, energy recovered) and only 33 per cent is recycled.

From stone to concrete

There have also been numerous innovations in building materials and processes. Resources that were ultimately non-renewable were also used from early on, from mammoth tusks to glyptodon shells, the latter from a giant South American armadillo that disappeared sometime after the arrival of the first hunters. With increasing urbanization, the need for building materials has become enormous.

With the exception of wood, there should be no global shortage for most building materials: limestone, sand, stone and clay are present in very large quantities that are a priori difficult to exhaust on a human scale, despite the very extensive uses that we make of them. But the main issue with these products, with the exception of some premium materials such as marble, is the cost of transporting them (they are of low value and high weight, so transport costs are a significant part of the total price).

There may therefore be real local shortages, in particular if some special characteristics are needed. For example, Dubai imports sand from Australia(!) because the wind-formed sand of its deserts is too smooth for use in construction projects. Closer to home, river sand, called 'rolled', is starting to be unavailable in the Paris region, partly because of over-exploitation, but also because taking sand from rivers is now limited by the need to protect wetlands, urban areas and so on. Crushing makes it possible to obtain sand from rock when rivers can no longer provide it (requiring a little more energy, of course), but such materials do not have the same mechanical characteristics and rolled sand remains necessary for high-quality concrete. We therefore either go further afield, at greater cost, to obtain materials, or 'ecological' considerations are undermined, with pressure to dredge the seabed or to relax local regulations … on pain of not being able to produce concrete satisfactorily or economically.

Finally, although sand as such remains widely available at a global level (silicon makes up 27 per cent of the Earth's crust), human activities are seriously disrupting natural cycles. Fifteen billion tonnes of sand are used in construction each year, dredged from rivers and the seabed, and more dams on rivers and water courses block upstream sediments. The source of renewal being cut off, many of the world's beaches are being eroded to the point where there is a risk that they will disappear by the end of the century.

Walls, structures, roads

Construction materials evolved little from ancient times to the end of the 18th century: large blocks of stone held together with iron bars sealed with molten lead, lime used as a hydraulic binder for mortars, wood as a structural material.

Lime is obtained by the calcination of limestone at high temperature, and transforms back into calcium carbonate after 'drying' in reaction with the CO_2 present in the air. It has been known since antiquity, especially by the Greeks and Romans, who understood that, by adding clay, sand, pozzolana (ashes from Santorini or Vesuvius) and even fragments of bricks, tiles or pottery, they could obtain a very solid mortar, which explains the exceptional longevity of some of their buildings. Italian pozzolans continued to be exported for many years to be added to locally produced lime, before the properties of the different constituents were identified and 'Portland' cement, patented in 1824, could be developed. This was a mixture of limestone and clay, but still quite a long way from the cements used today. Throughout the 19th century the technologies evolved (control of proportions of constituents, continuously fired furnaces, grinding of clinker and more) to improve and standardize performance and reduce energy consumption and wastage.

At the same time, blast furnaces and other processes were being developed in the iron and steel industry. In particular, the Bessemer converter, invented in the 1850s, made it possible to convert pig iron to steel in a very efficient way, opening the door to steel construction and then to reinforced concrete. Roads, formerly cobbled, were macadamized (paved with successive layers of increasingly fine materials) and then coated with tar (to give 'tar macadam' or 'tarmac') or bitumen (also known as asphalt, a by-product of petroleum refining), as early as the end of the 19th century in many places.

Wood

Forest over-exploitation is well known and characteristic of the difficulties in managing a theoretically renewable resource. For example, France is rather well endowed with woods and forests, which cover a third of its territory, and yet when we note that only 0.2 per cent of these are 'ancient' or 'natural', we can guess the pressure that was exerted on them in past centuries.

Deforestation was massive, early and irreversible, from ancient times in the case of the Mediterranean forest. It must be said that our ancestors did not hold back, what with Solomon and his temple constructed from centuries old cedars of Lebanon, Xerxes assembling a bridge of wooden boats to cross the Bosporus with his invading army, Athens and Sparta, Rome and Carthage, Mark Antony and Octavian and their consumption of many triremes in their numerous naval battles. After the reduction in the rate of exploitation that accompanied the fall of the Roman Empire, the great forests of Western Europe were gradually cleared, exploited as timber for construction (it took around 3,000 centenarian oaks to build a warship), and burned for energy for domestic use and for the fledgling industries (brick, glass, metals).

Wood has long been a standard material on construction sites (and still is for some scaffolding, formwork for concrete and so on), even for 'new' technologies. Thus, the expansion of the railways in the 19th century was as much based on wood technology as iron. In the United States in 1900, a fifth to a quarter of the trees cut down were used for railway sleepers, which had to be renewed (much too) regularly. American forests were probably saved by the use of creosote, a derivative of coal tar, to protect sleepers from rot and to reduce their turnover rate, and by the expansion of the automobile that led to the closure of railway lines.[13]

Here again, all strategies have been used to try to cope with local shortages: 'frontiers' were pushed back with the exploitation of more and more remote forests (especially for luxury species); innovations were used such as the chainsaw – which made it possible to fell a tree one hundred to a thousand times faster than with an axe – and mechanization of transport; substitution of timber including for example the use of aluminium and PVC for window frames.

Food production and storage

Arable land is also, if one does not take care, a non-renewable resource, because it can be eroded, exhausted or damaged by salinization. There are many historical examples, for example the consequences of the activities of civilizations between the Tigris and the Euphrates (nowadays rather bleak in agricultural terms), and around the Mediterranean (Libya was, with Sicily, one of the great granaries of Rome), or the enormous losses of arable land in the 1930s in the United States with the dust bowl,[14] the repeated dust storms that followed over-exploitation of the Great Plains' soils.

The constant struggle of static farmers to maintain and increase the productivity of their soils led to major technical developments, some very ancient:[15] the Roman triple of *hortus* (vegetable garden), *ager* (grain field) and *saltus/silva* (grazing and forest), in which animals were taken back into pens at night to collect their manure; the little-known agricultural revolution of the Middle Ages, based on the heavy plough, the working horse in harness and stables; the first agricultural revolution of modern times with fallow systems based on alternating cereals and forage (clover, alfalfa), then mechanization and the use of artificial fertilizers. By increasing the yield per hectare, this last development also responded to the shortage of arable land, but at the cost of unprecedented environmental consequences: eutrophication of the rivers, biological death of soils, emission of greenhouse gases (see Part III).

To complement the use of manure from livestock, many sources of fertilizers and soil-enhancing materials were used: human excreta from cities, dried in the open when carts came to empty building cesspits; urban sludge, algae, bones (rich in phosphates) from whales or even sometimes Pleistocene cave bears, then guano and saltpetre from Chile, brought back by the great Cape Horner ships of the end of the 19th century and exhausted in a few decades. As nitrates are used for both explosives and fertilizers, active research to synthesize them from the nitrogen in the air led to the development, at the beginning of the 20th century, of the Haber-Bosch process. Germany could then maintain its shell production during the First World War, even without access to Chilean saltpetre. After the war, agricultural practices changed radically, with Fritz Haber receiving the Nobel Prize for having 'saved' humanity from famine.

Finally, food preservation techniques helped in the response to shortages, either local or temporary, through transportation or storage constraints. With increasing urbanization, it was necessary to supply cities from the countryside. The transport possibilities between the localities of production and consumption played a major and sometimes limiting role. Paris, which was already populous from the early Middle Ages, owed its urban prosperity to an exceptional geographic location, well fed by a rich agricultural basin, itself served by a dense network of navigable rivers: the Seine, Marne, Oise, Aube, Yonne ...

The traditional methods of preservation, especially of meat, of which there are many (smoking, drying, pickling, salting, jam-making and so on), had led to the world being searched for spices for several centuries; spices were the driving force of the trade with the Orient and the exploration of the New World. These made it possible to eat hung (aged) meat and meat pies, but they were supplemented by two almost parallel 'revolutions'.

First, driven by the needs of the military, and by navies in particular, food packaging techniques made great progress at the beginning of the 19th century, after the invention, in 1795, by French confectioner Nicolas Appert, of airtight food preservation containers. Up to that point the method used was to store salted beef in barrels. Appert first used glass containers, and then metal came rapidly to dominate, using tinplate, that is to say tin-plated steel. Its use expanded enormously (for example, for the well-known corned beef), first with the American Civil War and then with the First World War. Today, we use more than 80 billion cans per year in the world.

In parallel, thermodynamics, initially the science of 'heat engines' – steam engines – led to the invention of refrigeration (which replaced the traditional use of blocks of ice, cut from frozen lakes and rivers in winter and stored) and freezing of food, which allowed its transportation over long distances. Refrigerated storage and shipping developed in the years 1860 to 1870.

These two techniques ensured the supply of meat to populations whose land could not have supported such a burden of livestock. Argentina, Australia and the Great Plains of the United States would in this way supply the burgeoning populations of Europe. The necessary supply of Europe by the United States during the First World War would accelerate the equipping of merchant fleets with refrigerated holds.

Change of scale with containerization

Globalization of markets was established by the steamship and railways in the 19th century and then in the 20th century by the diesel-powered ship and the truck. Transport played a big role, sometimes helping to overcome local shortages. But everything is a question of scale.

A huge step was taken with the introduction of containerization in shipping. Malcolm McLean, founder of the shipping company Sea-Land Service at the end of the 1950s, modified some liberty ships and tankers from the Second World War as container carriers to operate on North–South routes on the eastern seaboard of the United States. This revolutionary approach, reducing the downtime of ships in port (unloading containers by crane is much faster than traditional stevedoring approaches) and reducing the risk of theft, damage and loss (a few crates of whisky or other goods might be 'lost' from time to time to supplement dockers' incomes) can drastically reduce logistical costs, increase tonnages and carrying capacity. It allowed the Vietnam War to be fought (and all others since): without containers its logistics became unmanageable from 1966 owing to the number of soldiers engaged, and to their demands for large quantities of frozen meat, ice cream and other military needs. On their return voyages, empty ships stopped to reload in Tokyo and Yokohama, launching the first wave of Japanese exports to the United States in the 1970s.[16]

Containerization has thus been at the root of the explosion in trade in the past 50 years. It made possible the transport of finished or semi-finished products rather than raw materials (which had been transported for a long time, local resources often having been exhausted), and the redesign of logistical flows of large industrial groups. Of course, we were already transporting some finished or semi-finished goods: the East India Company shipped boxes of porcelain to Europe where it was much prized, and it also served as ballast around the valuable cargo of tea. But given the slow pace and cost of transport, it was used mainly when local production was not possible (for example, transport of iron or copper) or through the application of colonial economic policies, such as the export of British cotton goods which killed off the Indian textile industry in the first half of the 19th century.

Extremely low transport costs, thanks to containerization and the abundance of oil, are also, in a way, a 'solution' to local pollution, because they allow us, at an unprecedented scale, to put a distance between our actions (consumption) and their environmental and social consequences (from production). Pollution is outsourced to Bangladesh, China or elsewhere, just as out-of-town electricity and gas plants helped to move pollution out of cities in the last quarter of the 19th century. Edison allowed heating and lighting without the smell and soot of coal, oil or gas. Yet pollution is still there – coal plants remain by far the world's largest source of electricity and heat – but are located outside of dense urban areas. Electric or hydrogen-fuelled vehicles come with this same myth attached: they are 'clean' because they are odourless and they produce no harmful discharges in use. This is false, because it is necessary to produce electricity or hydrogen and, even if the energy could be easily available and truly clean, which is never the case, the manufacture of the vehicles, batteries, tyres, interior equipment and so on will always generate some unmanageable waste.

What would our European countryside look like, had it been necessary in recent years to build – and accept the discharges from – the new factories needed for our exponential consumption of telephones, computers, televisions, toys, clothes and chemicals? The answer is probably to be found in the landscapes of recently developed Chinese industrial areas.

The great turning point

So, even if there had been regular episodes of technical progress since Neolithic times or antiquity, it is clear that the crucial period from the late 18th to the end of the 19th century was decisive. The change in scale of production, the number and importance of technological breakthroughs, especially for manufacturing and for access to raw materials, mark the incredible acceleration of the fossil-fuelled Industrial Revolution.

Of course, we must not forget the 20th century, with its huge gains in productivity through mechanization, robotization and computerization. Not only have we discovered techniques to access abundant resources, but we have been able dramatically to reduce the amount of human effort involved in their production, moving from the coalminers of Zola's *Germinal* to ultra-mechanized Australian mines, from craft workshops to automated production lines, from hand-operated to numerically controlled machine tools. This is what has allowed the current mismanagement of material use and of the consumption of finished products.

Of course, this is not to reduce the Industrial Revolution simply to its technical dimension, or to the exploits of some inventors, even if they were geniuses. Because it resulted just as much, if not more, in social disruption, in the *great transformation*[17] of economic relationships: the concentration of work in factories and development of wage employment; the appearance of the 'free market'; the modification of property rights. Nor do we wish to idolize our 'miracle-working' engineers as Good Samaritans fighting against shortages faced by their fellow citizens (Thomas Edison, founder of General Electric, was not a choirboy but a shrewd businessman), nor to overstate the individual genius aspect of technical innovations.[18]

Nevertheless, the fact is that, overall, the entire technical system, built up and embedded in a social, moral and cultural system which it itself modified, has until now responded rather well to the risks of resource scarcity. But all this had a price, of course. That of a wild rush forwards, a permanent dashing from shortages to the new solutions to address them, that in turn create new shortages and new impacts. Among those are unprecedented pollution, social upheaval and environmental destruction. Engineers rarely make omelettes without breaking eggs.

Why high tech is not an answer this time

So why is it different this time, when the number of researchers, scientists, academics, laboratories, knowledge-sharing tools, 'industrial research partnerships' and exhortations to innovation have never been so high? Any rational mind should first of all note that because something has always happened does not mean it will necessarily continue to do so. The past does not determine the future, except for mathematical functions. If I jump from the top of a building, the fact that I arrive safely at the third floor will not stop me smashing into the ground. As the fool putting another coin into a vending machine says, "Don't stop me, can't you see I am winning!"

Of course, you tell me that we have heard it all before, that enough is enough: from Malthus to Dennis Meadows and his report to the Club of Rome in 1972, from 19th-century alarmists crying wolf about coal or ore reserves to the recent story of peak oil – we hear a lot about it but don't see much evidence. Certainly, gasoline and diesel prices rise, but quite frankly not much more than everything else: rent, property costs, the price of bread or a cup of coffee – especially in Europe since the changeover to the euro don't you think? And, if we compare the changes since the 1970s in the minimum hourly wage and the price of a litre of fuel, clearly the fuel is much cheaper today: you have to work less time to refuel your car, and fuel consumption may be better – but on the other hand urban development and our transportation habits mean that we drive more.

For the defenders of technology, everything looks pretty much set: for energy we will end up with nuclear fusion, or at least the fast breeder reactor will solve the problem of limited reserves of uranium 235. For the more 'green' among us, it will be clean renewables and hydrogen aplenty for energy storage and cars. And, meanwhile, the pressurized water reactors of third-generation nuclear plants and shale gas and oil will give us enough time to develop clean and sustainable technologies. To solve climate change, we will capture and store CO_2. As for the issue of metals, it is simple, since they are infinitely recyclable, long live the 'circular economy'. And for global food production, that will be addressed by genetically modified plants, together with aquaculture to deal with the collapse of fish stocks.

Sadly, this idyllic vision is based on a serious misunderstanding of several physical phenomena and of the 'systemic' nature of our industrial society, together with a certain understandable optimism when facing the difficulties that we have to overcome. Let us look at these issues a little more closely.

Quality and accessibility of resources: energy efficiency and 'peak everything'

Another truism: our industrial society is now largely based on the exploitation of non-renewable resources. To simplify, let's assume it is fossil fuels and metals.

Fossil fuels provide us with almost 85 per cent of our primary energy (excluding non-commercial biomass; see Figure 1.1), while renewables represent only 11 per cent (of which three fifths is from hydropower) and nuclear the final 4 per cent. Metals are available in limited quantities and are geographically poorly distributed, with the notable exception of iron and aluminium, whose quantities in the earth's crust are very large, 5 per cent and 8 per cent, respectively, and whose exploitation is above all an energy problem.

So, will our technological capabilities, current and future, be able to cope, in the short, medium or long term, with the risks of shortages in these two categories of resources, or will they not? That is, will we be able to transform potentially exploitable *resources* into *reserves* that are actually exploitable at the level of technology available and at market prices? By definition, reserves can be increased by finding new resources through exploration, by improving production and production techniques, or by raising the price. In the first two cases, technical and scientific knowledge is indeed very useful.

Whether for fossil energy or for most metals, the potential resources still underground (or underwater) are enormous: oil and shale gas of course, but also underwater polymetallic nodules, even methane hydrates and cobalt-rich deposits. We have hypothetical stocks that will allow exploitation for decades, even centuries (no one raises the question of beyond that, even if there are strange moral implications regarding the consumption of a finite resource, but, oh well). But there is a problem of quality and accessibility of these resources, because we have understandably started by tapping the richest, most concentrated, most easily exploitable resources.

If you interview metals specialists, most will tell you that there is no fundamental problem. Certainly, ore quality – in terms of metal concentration – is sharply down (for copper, lead, zinc ...), with those large mines of high metal content ores exhausted. But, 'all things being equal', it will be OK if we dig deeper, to extract more but lower quality ore, so there is no risk of scarcity. But it will be necessary to spend more energy per tonne of metal produced, and that is where the trouble begins.

Quality of energy resources

In the case of energy specialists, the answer is more ambiguous, since an additional factor must be taken into account. One can afford to mine an ore containing only 3 or 5 grams of gold per tonne, it is only a question of putting in enough energy to extract, to haul, to grind and to process the tonne in question to arrive at the price of the precious metal. It is different for energy. Here, you have to make sure you get a lot more energy than was invested in its extraction or generation, since the goal is to 'produce' energy. This is the measure of the effectiveness of this production or Energy Return on Energy Invested (EROEI). And, as it happens, this varies greatly depending on the resource (see Figure 1.2).

This is especially true for oil. Whereas in the 1930s we needed to 'invest' only 2 or 3 barrels of oil to produce 100 barrels in the giant onshore fields of Saudi Arabia (a return of around 40), today we must invest 10 to 20 barrels to produce the same quantity from an offshore field. We find the astonishing ratio of 1 barrel needed to produce 3 for Athabasca's tar sands in Canada, where a large quantity of the invested energy is in the form of natural gas used to heat the sand to extract and

Figure 1.2: Energy return on energy invested

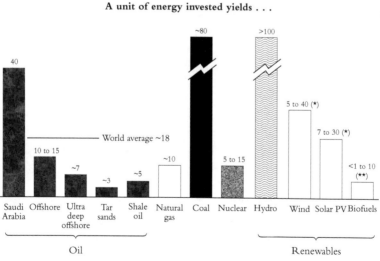

A unit of energy invested yields . . .

Notes:

* Function of technology and installation location.

** <1 for maize ethanol, 2–3 for temperate zone bioethanol (×2 or 3 in tropical regions).

Source: David J. Murphy and Charles A.S. Hall, 'Energy return on investment, peak oil, and the end of economic growth', *Annals of the New York Academy of Sciences* 1219(1), 2011

liquefy the oil (bitumen) that it contains. To be clear, we are burning gas to produce two to three times the amount of oil.

The resources of coal and natural gas are larger, sadly for the climate of the planet, but here too peak production will eventually be reached, perhaps in the decades 2020 or 2030, we'll see. This could even accelerate, as the scarcity of oil, which is a unique and difficult to substitute source of energy in transport (see Figure 1.3), leads to pressure to produce liquid fuels from coal or gas through Coal-To-Liquid (CTL) or Gas-to-Liquid (GTL) processes. This is the Fischer-Tropsch process, devised in the 1920s and developed by Germany during the Second World War to produce gasoline for the Luftwaffe from coal. Later, oil-embargoed South Africa became the world's leading user of the process.

Figure 1.3: Utilization of primary energy by sector

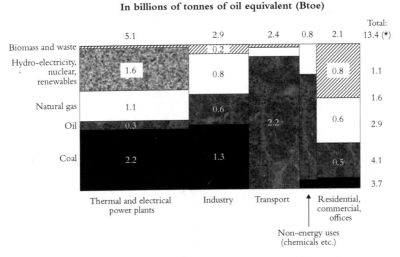

In billions of tonnes of oil equivalent (Btoe)

Note: ★ Of which 12.3 Btoe is commercial energy and 1.1 Btoe wood/biomass/waste.
Source: International Energy Agency (analysis by the author)

From peak oil to peak everything

To return to the issue of metals, less accessible fossil fuels also lead to an increased need for metals: compare a simple Texan oil well with its deep-water offshore counterpart, a gigantic steel platform surrounded by a swarm of supply boats, helicopters, high-tech directional drilling and so on. It is also bad news for renewable energies, which rely heavily on metallic resources, including rare materials such as neodymium and dysprosium in permanent magnets for wind turbine generators, gallium, indium, cadmium and tellurium for high efficiency photovoltaic panels (CIGS or Cd-Te), and copper that is used in greater quantities for every unit of electricity produced …

We could deal with constraints for one or other of the resources, energy or metals. But the challenge is that we now have to deal with them both at the same time: more energy is needed for less concentrated metals; more metals are needed for less accessible energy. Peak oil will likely be accompanied or followed by 'peak everything'.[19] And, to disappoint the most optimistic, we will not find our future metallic resources on the moon or in asteroids because the energy expenditure to get there is simply unacceptable, while the idea of thousands of years of nuclear energy does not hold water since it is necessary to rebuild power stations once or twice a century, without it being possible to recycle the materials that have been irradiated. Even with breeder reactors, it is necessary to resist the temperature, the pressure, the heat, the corrosion, the irradiation. In a few centuries or millennia, how will we do without the nickel, titanium, cobalt, or tantalum high-performance alloys used in nuclear plants? Without zirconium cladding for nuclear fuel rods? Without lead to absorb radiation and without tungsten for nuclear fuel containers? Without hafnium, cadmium, indium, silver, selenium or boron to absorb neutrons in control rods?

These geological and energy peaks are likely to be magnified by system-wide peaks, as other factors will exacerbate the purely physical and technical issues: problems of secure access to resources in unstable regions, lack of availability of personnel, for example experienced geologists (because the profession has had little recruitment in 20 years), issues in managing the growth of demand; ups and downs in the economic and financial system deterring investors, as new mining projects may require billions of dollars.

Limits of the circular economy

Of course, there is another difference between energy and metals: the latter, once extracted, do not go up in smoke like fossil fuels. One could therefore envisage recycling them indefinitely, once the quantity necessary for our society has been extracted (but how much might this be?). Of course, in reality, by virtue of the second law of thermodynamics, we always lose a little, whether at the time of recycling itself (loss in smelting) or during use (coins wear imperceptibly over time, because 'iron and brass are worn out with constant handling').[20]

Unfortunately, there are physical, technical and societal limits to recycling in a world as technical as ours. First, some materials, such as thermosetting polymers (polyurethanes for example), simply cannot be re-melted. Others, such as food or medical packaging, are not reusable once soiled.

Furthermore, the complexity of products, in terms of components (dozens of different parts in a mobile phone or a computer) and materials (thousands of different metal alloys, mixtures of plastics and additives, composite materials) prevents us from easily identifying, separating and recovering raw materials. For example, nickel, although easily identifiable (used in stainless steels among other things) and quite expensive, only has a recycling rate of about 55 per cent. Some 15 per cent is satisfactorily collected and recycled, but is lost in functional terms or is suitable only for less demanding use ('downcycling') because it has been contaminated or embedded in low-grade carbon steel, while ~30 per cent is lost in landfill and incineration. In three cycles of use, we lose about 80 per cent of the resource. And nickel is a fairly well recycled metal: for most 'minor' metals the recovery percentage does not exceed 25 per cent, and can be even less than 1 per cent for many 'high-tech metals' such as indium, gallium, germanium, selenium or rare earths.

Finally, a not insignificant quantity of metal is used dispersively, and is therefore not recyclable. It is used as pigments in inks and paints, as additives in glass and plastics, as fertilizers and pesticides (see Table 1.1). In this regard, some applications verge on the absurd: for example, silver

Table 1.1: Dispersive uses of some metals

Metal	% dispersive use (chemical excl. catalysis)	Examples of applications
Cadmium	~20%	Pigments (plastics, glass, ceramics), stabilizers (plastics)
Chrome	~10%	Pigments, tanning of leather
Cobalt	~15%	Pigments, tyres, adhesives, soaps, desiccants
Tin	~14%	Stabilizers (PVC), anti-algal paints, preservation of wood
Manganese	10%	Batteries, additives (ceramics), pharmaceutical purification (water)
Molybdenum	~30%	Pigments, lubricants
Lead	~12%	Additives (glass/crystal, ceramics) stabilizers (PVC), pigments
Titanium	~95%	Pigments, cosmetics, toothpastes
Zinc	~6%	Pigments, tyres, cosmetics, pharmaceuticals

Sources: Société Française de Chimie (French Society of Chemistry), Bureau de Recherches Géologiques et Minières (French mineral survey – BRGM), Mineral Info (http://www. mineralinfo.fr/ – French non-energy mineral resources portfolio), Industrial federations

– with its antibacterial properties – is used in nanometric form in socks, as an anti-odour technology!

Loss by dispersion (in production and use), through being discarded (for example, the tin can, the staple and the pen left in landfill), by reduction of function (by inefficient recycling), by entropic (marginal) loss: this is our destiny, the virtuous circle of recycling is being broken everywhere, and, at each 'cycle' of consumption, some resources are permanently lost. We are not going to scrape from old boats the anti-corrosion paint containing tin and copper. We will not collect from asphalt the particles of zinc or cobalt that come from tyre wear or from the platinum released in tiny amounts by vehicle catalytic converters. And we do not know how to recover all the metals present, in minute quantities, on an electronic card – hence the awfully low recycling rates of some of them.

Raising recycling rates is therefore a very complicated business, and does not simply involve collecting end-of-life products and integrating them into a processing chain. In many cases, it is necessary to review thoroughly the actual design of the objects, both in terms of the components used – the electronic circuit board with dozens of different metals present – and the raw materials themselves: mixtures of organic and inorganic compounds, such as in most plastics, printed objects such as food cans …

The 'green growth' craze

But the technologies we hope will save us are just adding to the challenges. By relying on technological solutions in our fight against climate change, we are likely to create new shortages (themselves requiring increased use of energy) and thereby drive the system in an unintended direction. This is because what we call 'green technologies' are generally based on new technologies, using less widespread metals and often with greater complexity in the products, thus increasing the difficulty of recycling. Let us take a few examples.

To reduce CO_2 emissions by a few grams per kilometre, without sacrificing either size or performance (principally the speed and impact resistance) of vehicles, the only solution, apart from increasing the efficiency of the engine itself, is to make the vehicle lighter. For this, we use high-performance steels, always more complex, alloyed with small amounts of non-ferrous metals (manganese, vanadium, niobium, titanium …). Not only are these no longer recoverable at the end of life (and the steels are therefore downcycled to carbon steel for use in reinforced concrete in construction, for example), but the expected level of 'purity' of the alloys is such that it is usually necessary to use virgin (first-melt) steels.

27

Low-energy or energy-efficient buildings also consume many scarce resources: equipment filled with electronics (sensors, building controls, micromotors for automatic electric blinds and so on), additives in low-emissivity glasses and more.

To develop sufficient renewable energy supplies to avoid being threatened by supply interruptions, it will be necessary to connect thousands of wind turbines, photovoltaic 'solar farms' and storage devices (in vehicle batteries, in the form of methane, hydrogen and so on) using smart grids (intelligent grids) to allow the balance between an erratic and intermittent supply and a variable demand. Consumers will also need to be connected by smart, communicating meters that will allow demand to be adjusted by load shedding. Such a technical macrosystem will be based on large quantities of high-tech equipment, stuffed with electronics and rare metals.

Scale and rebound effects

The problem of scale

Of course, interesting technical innovations may be introduced. There may be new products, such as better insulation for homes, or industrial processes that are more energy efficient or less polluting. However, there is a problem of scale, of 'roll-out': how to ensure the replacement of the existing and the widespread deployment of the new technologies in an appropriate time scale?

It takes 10 to 20 years to renew the fleet of automobiles in use, and therefore to bring them to the latest standard. In buildings, it will take decades, perhaps even a century, even at an accelerated pace of urban renewal and thermal insulation, to arrive at an acceptable level of energy consumption in the entire existing stock of buildings. Furthermore, there are many buildings (often part of our historic heritage) that will never reach acceptable levels, because they were not designed for good thermal performance in the first place, and maybe all that can be done is to give up heating them up properly!

As for more efficient industrial processes, there is the question of the book value of existing facilities, which must be amortized before being modified. Two examples are particularly telling.

'Conventional' coal-fired power plants have a thermal efficiency of about 35–40 per cent (for conversion of coal energy into electricity). For several years, there have been so-called supercritical or ultra-supercritical power plants, whose efficiency rises to 45 or even 50 per cent (a literally huge gain). However, they are more expensive to build so there may not necessarily be the incentive to install them. Most recent Chinese plants are therefore of the conventional type (in boom years, such as 2007–08, China built one plant per week, or the generating capacity of France, including nuclear, each year). Given the lifespan of these plants, we should not expect a gain in efficiency and reductions in CO_2 emissions for decades.

Some engineers and oil experts advocate the development of CO_2 capture and sequestration technology: the aim is to collect emissions from power plants and large factories (cement plants, blast furnaces) and to store them in saline aquifers or old oil and gas fields. In a delicious oxymoron, people talk of 'clean coal'. There are losses in the process (in the capture, transport, and compression of the gas), and potential risks, but this technology is seen by the highest international bodies as a defence against climate change, since demand for electricity and materials is growing and the use of coal is seen as unavoidable. However, given the necessary development time for the technology and the lifetime of each plant, it is likely that serious damage will be done before there is any significant effect.

The same goes for many innovations. The next time you come across an article showing a beautiful illustration of freight transport by sailing ship or airship, think of the thousands of container ships, ore carriers and tankers that criss-cross the oceans.

The example of renewable energy

Of course, we can, and must, develop renewable energies. Nevertheless, let us not imagine that they will be able to replace fossil fuels and allow us to maintain our current energetic debauchery. Yet, the promoters of the colossal projects that regularly appear in our major newspapers and magazines seem to think so.

An example is the Desertec project, which was proposing the installation, by 2050, of 20 concentrating solar plants (CSPs) in the Sahara, in order to produce 700 TWh (billion kWh)/year electrical output (~20% of *current* European electricity consumption). The cost for this was estimated at €400 billion, including high voltage lines. With an investment of about €0.5 billion per TWh of production (or 0.5 million Euros per GWh), it would make this type of electricity not so cheap: while today's best solar PV projects are still in a higher range (typically in 2018 from 0.5 to 1.5 million Euros of investment per GWh of annual production[21]), maintenance costs for CSP would certainly be much higher. As a result of the project's very uncertain economic viability, most industrial partners have finally withdrawn and it is operationally abandoned,[22] but the beautiful effects of the announcement have nonetheless remained in our collective consciousness.

Certainly, an area a few tens or hundreds of kilometres square in the Sahara or other deserts could provide all of the world's electricity, but these back-of-the-envelope calculations don't mean anything. To produce the world's 2018 consumption of 26,600 TWh it would be necessary to install the equivalent of 180 years of current production of solar panels (or, more modestly, 30 years for European electricity consumption)! Not to mention that these figures should be probably doubled or more to take into account storage losses due to intermittent production, and that after 40 years at most, we should start all over again, given the lifespan of photovoltaic panels. Moreover, who would sweep up around tens of thousands of square kilometres of panels after each sandstorm?

Furthermore, we must not confuse world *electricity* requirements and world *energy* requirements. I'll spare you the boring calculations (how not to confuse primary energy and final energy, how to compare the yields of heat engines and electric motors ...), but we would then need at least 600 years of solar panel production. Of course, I am being pessimistic, because we can achieve (green) growth in production capacity, for example the tenfold change that we have managed in the past, or even a hundredfold. Nevertheless, what an industrial challenge, so improbable! We would need to start building panel factories by the score, with huge supply chains ... not to mention the availability of metals, or the use of synthetic materials, derived from oil and difficult to recycle.

Another project, 'Wind Water and Sunlight', proposes to meet the energy needs of the whole world by 2050, using renewables only.[23] To achieve this, according to the 2017 scenario, it would be necessary to install 2.515 million 5MW wind turbines, 1.84 billion residential roof photovoltaic (PV) panels, 75 million commercial/government roof PV panels and 251,000 50 MW solar power plants among other devices (we admire the mathematical precision of the figures). Starting in 2015, nearly 12,600 GW of wind turbines would need to be installed in 35 – now 30 – years, despite knowing that we have installed at most about 60 GW per year in the last 5 years. It is simply a question of increasing the current production and installation capacity by a factor of 7 over the whole 30-year period. This would be enormously challenging, even if we ignore that steel, cement, polyurethane resins, rare earths and copper, boats and cranes are needed, among other things, to build and install a beautiful 'clean' wind turbine.[24]

Other celebrated technologies such as wave and tidal energy seem also quite improbable at large scale. The Pelamis system (a 150 metre long string of steel tubes producing 750 kW of power, or an annual output of around 2.7 GWh [that is, millions of kWh]), in order to produce the equivalent of a single nuclear reactor (8 TWh), required installation of the trifle of 3,000 devices, in 75 ocean farms of 130 hectares each. Future fishermen would have had to learn to weave in and out of the devices, if the company had been successful (it finally went bankrupt in 2014). Some other innovative solutions, such as tidal turbines, continue to be tested with public money, but we should be careful when betting on their real potential.

More credible scenarios do exist, of course, like that from the négaWatt Association[25] in France, whose great merit is to start by questioning energy demands and to propose a significant reduction in energy consumption by 2050. The production of 250 TWh in wind energy and 150 TWh in photovoltaic solar energy, in addition to a strong and systematic exploitation of biomass, remains however a very ambitious objective (compared to the current *world* production, of the order of 1,130 TWh and 450 TWh respectively using 2017 data).[26]

Let us summarize: wind, solar, biogas, biomass, biofuels, algae or modified bacteria, hydrogen, methanation, whatever the technologies, we will be caught out by physical constraints: by our inability to fully recycle materials (we make wind turbines and solar panels today using materials that we do not know how to recycle); by the availability of metals; by the land area that the technologies demand, and by intermittency and low yields.

The different renewable energy technologies do not necessarily pose a problem as such – no doubt it is better to have a wind turbine than a diesel generator with the same power. However, it is the scale that some people imagine they can achieve which is unrealistic. The widespread deployment of the right mix of renewable energies remains just a concept, and it will be difficult to respond to certain requirements. In particular, for the questions of transport and storage (and therefore the ability to adapt production to highly variable demands) there are no satisfactory answers, with an unrealistic need for metals (in batteries) or losses that are too high (for example, in gasification and methanation). There is not enough exploitable lithium on earth to power a fleet of several hundred million electric vehicles, and not enough exploitable platinum for an equivalent fleet of hydrogen vehicles.[27] Material constraints are seen everywhere in attempts to decarbonize.[28] And remember that hydrogen is not a source of energy, but only a storage mechanism. There is no doubt that fossil fuels, oil and especially coal, will be with us for some time. Even if that is not good news …

The rebound effect and growing needs

It is not just technical factors that constrain the impact of useful innovations or our ability to deploy them. Another well-known limitation is the *rebound effect*, in which the lower cost of use of something (owing to reduced energy consumption) increases demand and thus not all potential savings are achieved. For example, a car that consumes less fuel allows more miles to be driven for the same price, but use might be determined by budget. The rebound effect in this case would be 100 per cent (the improved efficiency leads to no reduction in fuel use). There can even be cases, as in the Jevons paradox, where the lowering of cost leads to an even greater increase in demand, as was the case with the introduction and improvement of the Watt steam engine.

The failure to achieve some of the economies that are theoretically available through innovation does not always come from consumers, but also from manufacturers. This is clear in the automotive sector, where there has been a steady increase in the efficiency of engines, which are then used to move heavier and heavier vehicles. What good is it to have a hybrid vehicle, if it weighs 1.6 or 1.8 tonnes empty?

Finally, it must be kept in mind that demand is growing, particularly as a result of the 'catching up' of emerging economies and the demographic effect of increasing population. While it is a good thing to be able to improve the efficiency of an industrial process by a few per cent or more, the consumption of coal grew 70 per cent between 2000 and 2018, and that of aluminium by 150 per cent. In one way or another, we will have to change gear.

Constraints brought on by innovation itself

It remains to be seen whether our current economic system and its need for growth, the operation and financing of research through public–private partnerships, the protection of intellectual property and the remuneration of patents will allow 'good' innovations to emerge that will save resources.

The answer is probably partly in the question. We know the need for physical or cultural obsolescence, driven by industrial oligopolies[29] and advertising, and it is unclear how things could change without an extraordinary regulatory thrust, which could be complicated to implement.

Some innovations, and perhaps those to be most feared because they are often ones that are profitable, allow the absurd needs generated by systems to be fulfilled. Nanoscale particles allow a sunscreen to offer protection and a feeling of freshness; RFID (radio-frequency identification) chips make it possible to pay for the contents of your supermarket trolley without going to the cash register; office glazing can be made opaque by electrical impulse; refrigerators prepare the shopping list. We are far from the 'revolutions' of the steam engine and crop rotation!

These examples are caricatures, but emblematic of many of today's 'innovations'. Frankly, would we develop such high-tech applications if we knew the real price to pay, in terms of factory waste, the disappearance of nature and resources, the destruction of everything that has sustained us, physically and intellectually, for millennia? We cannot say. However, since these innovations are often applications that are civilian spin-offs from military research, we don't ask too many questions before imposing them on ourselves. Innovation has always been intimately linked to military technology. The copper or bronze axe was more practical than the stone axe to clear vegetation, but also to smash the skull of an unpleasant neighbour. Today's developments are especially high tech – in aeronautics, the internet, electronics and telecommunications. The current convergence on nanotechnologies/biotechnologies/artificial intelligence/ cognitive sciences (NBICs), in applications from the infantryman of the future to miniaturized drones,[30] is the most recent example.

Systemic effects and decreasing returns

Everything is connected. Since our entry into the Anthropocene, the great leap forward of the Industrial Revolution, whenever we think we have found a technical solution to a shortage, because our Earth is finite, we often create other shortages or negative effects elsewhere. The different problems interact with each other, and are reinforced by 'positive feedback' loops that act as aggravating factors: less metals implies less energy and therefore less access to metals; urban sprawl increases the consumption of energy and metal resources; agricultural practices deplete soils such that they require more supplements, so more energy. Some innovations serve only to manage the complexity or negative effects of previous innovations.

What is the resilience of an increasingly complex and interdependent system operating in just-in-time mode? What is its ability to withstand future economic and social disruptions, which will no doubt be numerous and at a global scale? It is unquestionably very fragile today, because the industrial and financial system has become so interconnected and unmanageable. Large companies assemble subsystems manufactured around the globe, with raw materials sourced from dozens of countries.

The more complex we make systems – and without doubt more technology means more complexity – the more difficult it will be to maintain them economically and, potentially, the more sensitive they will be to external disturbances: climate change, resource scarcity, geopolitics, Fukushima-type industrial disasters or pandemics.

Some see in renewable energies a possibility of re-localization, of a community control of energy production. This is undoubtedly possible for simple technologies (domestic thermal solar or small wind turbines), but surely not for the high-tech developments that we are promised. The manufacture, installation and maintenance of the technical monsters that are wind turbines of 3 or 5 MW are only within reach of a handful of transnational companies that rely on a global production organization and highly centralized expertise, implementing expensive industrial practices (engineering, design, advanced materials, computer simulations, extraordinary logistics – barges, boats, cranes, special trailers …). Future customers will depend on an extensive supply network for the manufacture and distribution of cutting-edge spare parts, stuffed with complex power electronics. This is far from autonomous, resilient, community-based production, controllable by local people and companies.

To finish, the latest pet high-tech themes

Given that calling for an increase in the rate of recycling will not be enough, what are the latest promises that our scientists and miracle-working engineers are making? To respond to material shortages, the three winners are predicted to be, first, replacement of fossil resources by the 'bio-economy' for non-energy use of oil, gas or coal through use of renewable 'bio-resources', saving or substituting for rare raw materials; second, nanotechnologies exploiting miniaturization and, in particular, manipulating materials at atomic scale; and, third, the dematerialization of the economy allowing a future 'Economy 2.0' to grow without increasing demand for resources, or even effecting a reduction.

The bio-economy

So, what are we talking about when we say bio-economy? Table 1.2 shows the 'marketing' colour codes used to describe biotechnologies in some institutional presentations or for informed audiences.[31] The codes were developed to distinguish different industrial applications of biotechnologies, for the sake of better societal acceptance, to differentiate for example between medical examples, which are difficult to object to, and GMOs, whose reputation is already somewhat tarnished.

'Green' biotechnologies for the food industry are the best known. They mainly involve genetically modified crops (GMCs; see Part III), but also animals, such as the latest AquAdvantage transgenic fast-growing salmon

Table 1.2: Biotechnology for dummies

Biotechnologies of 'colour'	Basic principle	Principal products expected as outputs	Industrial maturity
... green Agri-food	Genetic modification of existing plants and animals to adapt their properties	Genetically modified crops (GMCs) • Pest resistance (Bt gene) • Herbicide resistance • Properties or composition of modified foods Genetically modified animals • Fast-growing salmon	Already strong
... white Industrial applications	Use confined to micro-organisms or enzymes (existing, modified or synthetic) to produce intermediate chemicals from cultures, residues or wood	Intermediate products for basic chemistry (surfactants, lipids) Biofuels Biomaterials (polymers ...)	Starting up Developing rapidly
... yellow Treatment and elimination of pollution	Use of living organisms (plants, algae, micro-organisms) to degrade toxic wastes into non-dangerous products	Clean-up of water and soil Bio-leaching (mining industry)	Low
... red Pharmaceutical industry	Use of very diverse technologies	New treatments, vaccines, diagnostics ... New therapeutic molecules Gene therapy, cell therapy	Already strong Developing rapidly
... blue	A class sometimes used for products linked to marine biodiversity (algae etc.) Cf. green or white		

35

and, perhaps soon, pigs with lower levels of nitrates in their urine and kangaroo bacteria used to give cattle methane-free flatulence to save the climate. The term genetically modified organisms (GMOs) covers plants and animals. Essentially, they consist of tinkering with the genetic code of existing plants and animals to obtain characteristics that are commercially interesting: resistance to pests, herbicides and so on. To do this, we transgress the boundaries between different groups in nature – insects, mammals, bacteria – as in Bt plants, for example, that contain some genes of the *Bacillus thuringiensis* bacterium, or goats that produce milk with spider silk protein, thanks to an appropriate gene.

'White' biotechnologies encompass a number of industrial applications that aim to produce, from renewable resources and using biological processes, a great variety of molecules – intermediate chemicals – that are then transformed into chemical products (surfactants, lipids, lubricants, solvents, acids ...) and biomaterials ('bio-polymers' or 'bioplastics'). These can be used in many sectors: cosmetics, leather, paper, agri-food (sweeteners, preservatives, thickeners: *bon appetit!*), construction, textiles ... and also the production of biofuels. This is something of an about-turn, because, in the 1960s and 1970s, it was envisaged that we would feed ourselves using petroleum products. We should be cautious about what we are promised by our scientists and industrialists!

White biotechnologies employ, a priori in a closed environment (in industrial chemical reactors to avoid any leakage), genetically modified bacteria or enzymes obtained by 'site-directed mutagenesis' or, eventually, by synthetic biology (construction of all parts with the help of electrical and computing technologies). It is therefore, roughly speaking, basic fermentation, but with organisms whose genetic material has been worked on to optimize the output product. This is referred to as a cell factory, or 'enzymatic biocatalysis'.

The raw materials for this are typically derived from crops rich in an easily exploitable material such as starch, sugar (contained in corn, potatoes, beetroot ...) or oil (from oleaginous crops). There is of course potential competition with food uses, but no, biotechnologists have thought of everything and we could also equally well use crop residues such as straw or wood. One can even imagine whole plants being 'digested' in sorts of 'bio-refineries' that would produce many different molecules as outputs.

There are other colours. 'Red' is used for the well-established area of health, medical, diagnostic and pharmaceutical applications. 'Brown' is applied to desert and arid zone technologies and 'grey' to classical fermentation technologies. 'Yellow' is sometimes applied for uses that aim to treat or eliminate various pollutions, for example of water or soils. 'Blue' or marine biotechnology refers to the use of molecular biology or microbial ecology of marine organisms in health products, cosmetics and aquaculture, although some prefer to divide such techniques into white, green or red as appropriate. Thus, our transgenic salmon falls into the blue category, although some have classed this as green.

To summarize then, what is the purpose of the bio-economy or what service does it provide? It mainly involves three types of 'product'. First, genetically modified plants and animals, with the promise of 'feeding the world', to avoid food shortages for the extra billions of future populations (see Part III). Second, biomaterials from bio-resources (purposefully grown or from wood or plant residues) that will feed a 'plant chemistry' sector as oil and natural gas runs out or increases in price. Third, generations of biofuels – first, biodiesel/bioethanol, already established; second, based on the processing of plants or their residues; and, third, biodiesel or hydrogen from the cultivation of micro-algae – to fill the tanks of our road vehicles, aircraft and ships.

These technologies are quite effective at producing a few tons of aspartame from sugar or starch. But can they give us an alternative to fossil fuels, even if we ignore the risks of dispersion or biological contamination associated with GMOs?

The first issue is that the technologies are not sufficiently mature, depending on the type of biomass used as input (see Figure 1.4). So long as we have simple inputs, so that we can remain in 'mono-process' mode, the possibilities of application are not so distant. This is the case for oleochemistry – the chemistry of oils – which is relatively well understood and does not involve much in the way of biotechnology. Chemistry from sugar and starch is also feasible and already exists for simple processes and outputs. But things get difficult when we get to cellulose and hemicellulose, which make up the structure of plants, and even more so with lignin, the main molecule of wood. For the moment, its exploitation is only considered thermochemically, that is to say by gasification, at the cost of considerable energy loss. Seemingly, we are not yet managing to domesticate, like the termite, a bacterium that will allow us to 'eat' wood to turn it into useful chemicals.

Perhaps more serious issues arise when we try to get a mix of all the different molecules in a biorefinery using only crop residues. Then the level of complexity will increase exponentially. Without pre-judging the abilities of our miracle-working engineers, it is a safe bet that we will start by exploiting crops at the expense of food uses, probably also using GMCs to increase the content of useful molecules in the plants (for example consider the BASF Amflora transgenic potato, with starch composed almost exclusively of amylopectin). But these issues are nothing that cannot be solved, maybe, with ever more innovation.

Figure 1.4: White biotechnologies and plant chemistry

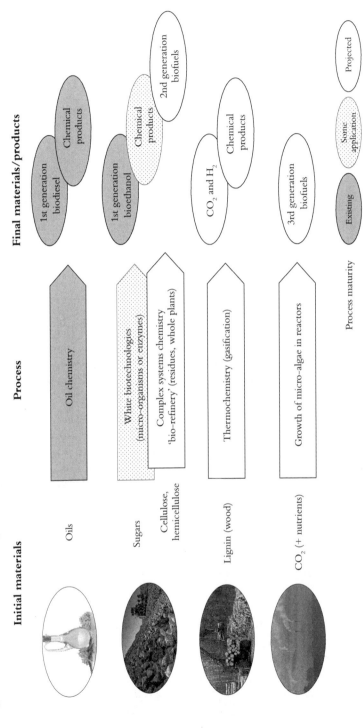

Initial materials

Process

Final materials/products

Oils

Sugars

Cellulose, hemicellulose

Lignin (wood)

CO_2 (+ nutrients)

Oil chemistry

White biotechnologies (micro-organisms or enzymes)

Complex systems chemistry 'bio-refinery' (residues, whole plants)

Thermochemistry (gasification)

Growth of micro-algae in reactors

Process maturity

1st generation biodiesel

Chemical products

1st generation bioethanol

Chemical products

2nd generation biofuels

CO_2 and H_2

Chemical products

3rd generation biofuels

Existing

Some application

Projected

In the second, more acute problem, technological factors are of secondary importance. We only need to compare the orders of magnitude involved to be convinced of the impossibility of 'bio-sourcing' our astronomical consumption of plastics, chemicals, lubricants and, more seriously, fuels. If we try, we are back in the 19th century before the invention of the chemistry of coal and gas! In the bio-sourcing of products we come across the age-old constraints of available land, and the limited possibilities of improving yields in crops and processes.

Admittedly, according to the promoters of plant chemistry, annual world biomass production is of the order of 170 billion tonnes per year of dry matter (75 per cent sugars, 20 per cent lignin, 5 per cent oils, fats, proteins and other materials), which may be compared to the very modest human needs: about 6 billion tons, of which a good half is for food. Well, maybe. In reality, the vast bulk of this biomass production is clearly not exploitable: first, owing to accessibility issues (from Arctic lichens to tropical forests, to the immensity of the oceans); second, because it is necessary to feed the different trophic levels of the food chain (50 billion tonnes of phytoplankton feeds zooplankton, then anchovies, before the fish of the third or fourth trophic level, which constitute the essence of our fisheries, are fed), and, finally, because most must return immediately to the natural cycle – via microbial decomposition – or the delicate balance of nutrient-limited ecosystems will be interrupted.

In order to convince ourselves of this, let us start with the current need for chemicals alone, which is more than 400 million tonnes of carbon content, or 9 per cent of oil production, 5 per cent of gas, 0.1 per cent of coal and 3 per cent of biomass. Let us compare it to what we could get from all of the world's biological cultures.

Excluding wood and fruits and vegetables, world agricultural production amounts to about 12 billion tonnes, including 5 billion tonnes of grains and other carbohydrates, and 7 billion tonnes of residues. Two thirds of this mass is not carbon (mainly oxygen, hydrogen and nitrogen). In 'carbon content' we are talking about 4 billion tonnes. To use this to make chemicals, it must first be transformed into substances that can be exploited by bacteria or enzymes, that is, into fermentable sugars (which is complicated for hemicellulose), which would give a yield in the order of maybe 50 per cent. Starting with a 60 per cent stoichiometric yield (from proportions in chemical reactions) and a reasonable industrial yield of 70 per cent, this would yield a little over 800 million tonnes of carbon in total (4 billion × 50% × 60% × 70%). But only 250 to 300 million of these are from residues because, except for rice, there is more carbon in the grain than in the residues.

Let us recap. Imagine that we could recover in the order of 100 or 200 million tonnes of carbon of the 250 to 300 million theoretically 'available'. This is pure utopia of course, because, at such a rate of recovery, how would the fertility of the earth be maintained through the necessary nutrient capture, not to mention the purely logistical issues in harvesting all the agricultural residues of the world? But our current need is already for 400 million tonnes! We will not be able to 'bio-source' our future polypropylene car bumpers and disposable nappies based on acrylic acid, far from it, even if we can increase yields a little or cultivated land by a great deal.

And remember, we are only talking about the need for raw materials for chemicals. If we were to add the need for lubricants and especially for liquid fuels, we would have to increase these figures by at least a factor of five (to more than 2 billion tonnes of carbon content). We don't need to calculate that even if we add wood lignin as a resource, the problem remains, even if we take a large chunk out of the world's forests. So we can forget about second-generation biofuels based on crop residues. '2G' is already outstripped, as in telecommunications, but there remain the '3G' biofuels, to be produced using micro-algae in huge production plants in the desert. But what consumption of metals and other materials is required to install, maintain and replace the millions of pipes needed for production at the scale of hundreds of millions of tonnes? Once more, we fall back into the problem of industrial deployment that we have already met.

Nanotechnologies

So, what about nanotechnologies? Here again we are promised all sorts of future wonders: 'smart' materials, materials for health and care, for environmental clean-up and surveillance, for production of drinking water and for renewable energy, all while opening up the possibilities of material savings and substitution by more abundant resources. In short, we can have our cake and eat it! Let us leave aside here the question of the very real risks related to the manufacture, use and dispersion of nanoparticles and concentrate just on the 'promises'.

Nanotechnology comprises broadly three activities. The first is 'nanobiology', with links to biotechnologies, hence the convergence of N (nanotech) and B (biotech) in the famous NBICs. The second is 'nanoelectronics', the worthy child of microelectronics at the nanoscale, which continues the historical trend of miniaturization of electronic equipment, taking advantage of the special characteristics of materials at this scale. The third is 'nanomaterials', which are of particular interest to us.

The ingredients of these nanomaterials can be carbon nanotubes, nanoparticles of clay or silica (SiO_2), metallic nanoparticles: mainly titanium dioxide (TiO_2), zinc oxide (ZnO), aluminium oxide (Al_2O_3), silver and gold, but also iron, cerium, tungsten, zirconium. These are then incorporated into textiles, ceramics, paints, coatings, varnishes, composite materials, glasses … or incorporated as additives in creams and gels, cements, and even foodstuffs!

The main applications today are in the 'health and fitness' sectors: tennis racquets, but especially cosmetics, sun creams or revitalizing, repairing creams, concealer for example, stuffed with titanium or zinc nanoparticles to better penetrate the skin – yuk! – where they release their active ingredients. The food sector is not being left out, with the use of silica-based nanoparticles as an anti-caking agent (E551 and others), the first applications in 'smart' packaging with better barrier strength or built-in freshness sensors, and artificial nano-encapsulated flavours and agents. If your sugar, salt or cocoa powder does not stick to the bottom of the container after some time, as in the past, you will know that you are being made to eat nanoscale sand without your knowledge.

Nano-silver is widely used (in several hundred products), for its antibacterial properties – in textiles, food packaging, refrigerator interiors or washing-machine drums. The plaster, paint and varnish sector has also found its first applications: anti-scratch paint, self-cleaning titanium dioxide or water-repellent coatings and so on. Most of the other applications are still at research and development stage, and we are still far from reasonable-cost photovoltaic paints or fuel cells becoming widespread.

From this rapid overview we can identify the technical limits of nanotechnologies: they are supposed to save material by promoting miniaturization (undoubtedly true for nanoelectronics) or substitution (again, in electronics, conductive carbon nanotubes can replace copper). However, in the vast majority of cases, applications are dispersive, which of course incorporates metal particles into products with no hope of recycling. This is not necessarily so serious when it comes to silica, but is much more annoying when it comes to zinc, titanium or silver. The volumes involved are not insignificant: the production of nano-silver is currently estimated at 500 to 1,000 tons per year, or about 3–5 per cent of world production of silver metal. This is called throwing money down the drain! And the number of applications is exploding.

In nanoelectronics, material consumption is potentially lower, for the same functionality, but possibilities of recycling of the numerous metal components are reduced – or at least not improved. We do not yet know how to achieve a satisfactory polymetallic recycling of microelectronic chips and circuits: when they are recovered and treated, we are often satisfied just to recover some profitable elements such as gold, or possibly copper, and some precious metals like platinoids.

Economy 2.0 and the dematerialization of the economy

Since neither biotechnology nor nanotechnology can meet the growing challenges of lack of raw materials, there are only opportunities to recycle (we have seen the limits of this) or to do 'more' (gross domestic product [GDP], commercial activities, 'wealth', services ...) with less, that is to say 'dematerialize' or reduce 'resource intensity' of the economy.

Historic mirages of dematerialization

Remember, not so long ago, the computer age was about to revolutionize our relationship to the material world. We would save money on paper because we would no longer need to print, and on travel, thanks to teleworking and videoconferences. Unfortunately, we have never used as much paper (between 2006 and 2017, world consumption of paper and cardboard increased by 10.5 per cent – we could do better for a dematerialization revolution), and we have never travelled so much for professional reasons.

We were lured, probably by a perverse rebound effect. We print a smaller percentage of the documents we see on our computers, but we receive more things and printing is easier to access (printers are lower cost and more reliable than in the past), so we print more in total. "Wait a moment, I'll print you a copy!" we hear everywhere. That's how 7 million tonnes a year are used in Europe, just for office paper. One of my old bosses, in response to my repeated complaints – I was then working in a sector (consulting) paid almost by the kilogram of paper embedded in its thick mission reports – told me that the area of forest was increasing in France. He was doing a good job of ignoring the global pulp market, with eucalyptus trees replacing the primary rainforest in the Amazon, high energy consumption and industrial wastes (of course he did not live next door), and many chemicals – a lot of them non-renewable – used in the production processes.

The IT and telecommunications sectors are of course not virtual, and nor is the internet itself: servers, wireless antennas, terminals, accessories, and transoceanic fibre optic bundles that we continue to install regularly, in line with the rise in traffic, all consume energy (about 10 per cent of worldwide electricity) and raw materials (see Part III). And, when we compare the environmental impact of 'good old paper' with the digital alternative, whether it is an e-reader, tablet or otherwise, it is almost certain that we have not gained from the change.

Towards a service economy

The service economy is a good idea in principle: by offering to rent/lease rather than buy items, we encourage the supplying company, that remains the owner, to properly maintain the equipment to make it last, and not to rely on planned obsolescence. The case of business photocopiers is often cited as an example of this 'good practice'. Looking closely, I'm not sure that it has achieved much, given the rate of technical changes in devices, such as the arrival of colour, integrated scanners and faxes, and increased speed.

There are also limitations in the model. It might work for important objects like cars, provided however that we can avoid the phenomena of fashion or the tyranny of marketing and the social standing afforded by a company car. Which company, for example, could afford to offer remanufactured 25-year-old cars to its senior employees?

It is the same in computing or telecommunications. It would be quite simple to rent computer terminals, but the obsolescence of products is today more symbolic than technical. It may work for big domestic appliances, that we use almost every day, but it is more difficult to envisage for small objects. Imagine having to pay every month for the use of a child's bike, a hair dryer or a radio alarm clock? And would such a system not threaten the least well-off households – a payment needed every month, and as soon as one is no longer in credit, the company comes to repossess the refrigerator?

And all that to make sure that refrigerators last 20 years as was the case in the past? Is there not a much simpler, regulatory route, such as the obligation for manufacturers to offer a long guarantee and repair and recovery of their products?[32]

3D printing

Some see 'fab labs' (FABrication LABoratories), and the well-known 3D printers that they use, as a means of re-localizing and democratizing production, allowing users to repair objects by manufacturing replacement parts, in short, the emergence of a true material Economy 2.0 and the end of Fordism and classic factories.

So, what is this all about? Fab labs are places that gather together digital manufacturing tools chosen for being relatively easy to access and use, such as milling machines, laser cutters or even sewing machines and, more likely than not, 3D printers. The concept is not without merit: it can effectively enable the design and production of small-scale products, by sharing expensive equipment between different people and micro-enterprises in one place.

But with respect to the supposedly revolutionary 3D printers the situation is different! These machines are generally based on Additive Layer Manufacturing (ALM) technology: the objects are produced by deposition, layer by layer, of a basic material. For this, the material, often a thermoplastic polymer, polyamide or polypropylene, but sometimes metal such as aluminium, stainless steel or titanium, is brought to melting temperature and deposited by the printer on the previous layer. Various processes are used, including extrusion of a molten filament, laser sintering of a powder, curing of a photo-sensitive material and so on. It is an additive technology, the opposite of machining which is inherently subtractive: drilling, threading, cutting and so on.

According to the media, this is 'the steam engine of the 21st century' and soon we will all be able to make car bodies and parts, batteries, but also custom-made shoes, guitars, pizzas, houses, even space stations! But if we look closely, it is about 'mono-material' objects, first and foremost in thermoplastic or thermosetting resin. This might work for car bodies, shoes or some types of houses which are composed of 'sheets' or 'panels' of single materials, but we could not make a piano or a phone with a 3D printer. To produce metal objects, it takes very special 'printers', which melt or sinter the metal by laser or electron beam, a technology unimaginable use by for private individuals. 'Consumer' 3D printers only work with polymers. We will not use them to make nails or screws.

Those who claim that classic production will soon disappear are therefore seriously mistaken. First, they confuse process industries with part manufacturing. Fab labs with 3D printers could, at best, replace the latter, but certainly not blast furnaces, cement plants, glass factories, or refineries and chemical plants that make the powders and filaments that serve as 3D 'ink'!

Second, they quickly forget that most of our products are not made of a single part or material, but instead are composite and need assembling, often by hand (hence the term *manu*facture). Massive high-tech electronics factories are still there for the moment, despite what futurists may say about the end of Fordism. 3D printers will not be able to print themselves.[33]

Finally, the only materials that can be processed are those that can be melted: fabrics will continue to be woven and sewn. Metals are suitable only if they do not have to undergo heat treatment or quenching (we will not be able to manufacture a beam or a high-performance ball bearing), glass if it is not tempered or laminated, etc.

There will ultimately be two kinds of applications of 3D printing. Some will be expensive, and generally reserved for advanced industries or high-value applications, such as for aerospace parts or dental implants. Others will allow the general public to make small plastic objects. Yes, you will be able to make a beautifully personalized mobile phone cover at home, or a new collar for your dog, but without the flea repellent, because the 3D printer cannot include a chemical plant. Undoubtedly there will be enough interest to raise some funds for start-ups, and enough to entertain some geeks, before these objects join the ice cream and bread makers in our cupboards and waste disposal facilities!

Industry 2.0, secondary or tertiary sector?

By ignoring the material reality, futurists are also mistaken when they speak of a world in which we will all be producers – of objects using 3D printers or of energy via the widespread installation of photovoltaic solar panels.

By producing energy from solar panels, by printing objects, I am not producing anything at all. I am rendering a service: first, with my property, having dedicated some area of my roof, and, second, financially, having invested in the solar panels. This activity is hardly comparable with that of a traditional carpenter, for example, who can choose a tree according to some criteria, and make a piece of furniture from it. Or a plumber who solders a joint. I do not exploit any know-how. I am only a tertiary outlet of the real producer, the manufacturer of solar panels that remains anchored in the good old Fordist secondary sector, nowadays often made in China or South-East Asia.

The triple dead-end of an extractivist, productivist and consumerist society

Green growth and new technologies therefore at best make it possible to slow down the collapse, and at worst will inadvertently speed it up, through systemic effects. Our technological society has gone down a triple dead-end, and extracting it will be difficult because there is so little room to manoeuvre.

The first dead-end is that of resources, either non-renewable or exploited at an unsustainable rate – fossil fuels, metals or minerals such as phosphorus, forests and fisheries. It also involves collapse of biodiversity and degradation or disappearance of agricultural soils. This happens through salinization following excessive irrigation – the same syndrome that caused the collapse of the Sumerian civilization between the Tigris and Euphrates (and already 5 per cent of the world arable land is affected) – and by biological destruction (for example, intensive use of fertilizers and pesticides) and erosion, with 20 per cent of soils significantly degraded.

Wind or water erosion attributable to anthropogenic action leaches 25 billion tonnes of soil per year worldwide, or 0.3 per cent of the total mass of arable land: which means that, broadly speaking, we will be left without arable land in about 300 years. Of course, this is only an average; some soils are in good condition while others will disappear in less than 10 years. In France, average erosion is 10 tonnes per hectare per year (up to 40 tonnes depending on the soil, and in some places, such as the beet-growing northern plains, it can easily be increased to 100 tonnes). Similar rates are seen in many other locations around the world. Producing a tonne of food also 'costs' us a tonne of soil (see Figure 1.5). And the world average is closer to 1 tonne of food for 4 tonnes of soil that disappear!

The second dead-end is related to pollution: of course, the over-production of greenhouse gases, but also the metals we extract from the lithosphere, the plastics and persistent organic pollutants that we synthesize, the dioxins produced by incineration – that contaminate soils, fresh waters, oceans and, ultimately, living things. Some of this pollution is of such magnitude that it is highly newsworthy, such as plastic debris

Figure 1.5: A tonne of soil for a tonne of food

French agricultural production

In millions of tonnes per year
(rounded figures – variable depending on the year)

Wheat	40
Maize	15
Barley	10
Other cereals	3
Rape	5
Sunflower	2
Other oilseeds	<1
Potatoes	5
Beets	35
Fruit and vegetables	8
Milk	25
Meat	5
Wine	5
Other production	<3

About 160 million tonnes per year

Erosion of agricultural soils

Areas of high or very high risk

About 150 million tonnes per year
10 tonnes per hectare per year (average)

Sources: IFEN, French Environment Institute, French Government Ministries, ADEME, French Environment and Energy Management Agency

in the North Pacific and North Atlantic. In other cases, it is progressing more slowly: soil pollution by metals from car exhausts, asphalt roads, slag (incinerator ash) used in road sub-layers; sewage sludge spread over fields.

The third dead-end is the consumption of land that we engulf with bitumen and concrete. Globalization consumes it through its growing and ever-changing need for distribution, logistics and transportation. Planned obsolescence affects objects but also locations. For example, for supermarkets, we are regularly creating new roads, which change travel times and cause shifts in catchment areas. Or it is the opening, a little further away, of a new shopping centre with a hypermarket and multiplex cinema, which attracts crowds and empties other hypermarkets. The lucky ones then turn into garden centres, others are left deserted but the bitumen remains. This is how, in France, urban development has taken place at the incredible rate of between 60,000 and 75,000 hectares per year for more than 20 years (and it has been accelerating in recent years). This is the area of a French *département*, 1 per cent of the total land area, in 7 to 10 years! Two thirds of this area is built upon or asphalted (roads, car parks), the rest transformed into roundabouts, motorway borders or golf courses, to the detriment of crops, often on the best land, meadows, moors and hedges.

In addition to this triple dead-end, which is physical, there is the social dead-end, with the widening of inequalities – although history has shown that we can go a long way in this regard without the system collapsing – and the moral dead-end. A world that fells ancient forests to make paper tissues or plywood, or, closer to us in Europe, a world that heats café terraces without being disturbed by this senseless waste, shows that, here too, the limits can be pushed quite far.

In inaccessible technical jargon we are told of the reasons for the 2007–08 financial crisis: problems of credit, indebtedness, confidence, exchange rates, currencies, credit ratings and so on. But any sensible person who has recently set foot in Spain, for example, will be only too aware of the house of cards on which growth fuelled by credit has been based: oversized highways, *hundreds* of airports, titanic works of art, empty buildings whose sole purpose was purchase and resale. How long could all this go on? And, especially, how might all this be set right?

The exhortations for the return of growth are pathetic. It will not be enough to rediscover the 'trust of markets' or of households, to be innovative, in a green way of course, or to print money to boost consumption (strike out whatever does not apply depending on your politics). Because we cannot concrete *ad vitam aeternam*, and the sought-after dynamism of countries like Dubai or Singapore is obviously and thankfully not reproducible.

To take a metaphor from the world of computers, it is time to change the software; but what I am saying is that it is also the motherboard and all the other circuits that we need to change.

PART II

The Principles of Simple Technologies

Let us summarize the first part of the book. We are concerned about the availability of resources to us in the near future. Our current approach, to push forward as fast as possible using high-tech innovations, will not solve the problems that we face. At best, there are some good ideas, but we will not be able to implement them at the speed and to the extent needed, far from it. At worst, the new technologies and their harmful consequences will only accelerate the issues that have driven us to this impasse. Between the two, pointless technological goodies may falsely delight naïve media, sustainable development advisers, economists and futurologists.

So, is this our destiny? Is a *Mad Max* style dystopian struggle for survival inevitable? Will we return to the Stone Age, to lighting by candle, to the Middle Ages of *Les Visiteurs*[1] (take your pick)? Before giving way to fatalistic notions that "anyway it is the fault of the Chinese and Indians who are too numerous" or the survivalist's "it's all going to blow up, I'll bury a box of assault rifles in my garden", let's see what other options are available to us.

Even if the demographic question cannot be avoided, it is evident that it does not help us to solve our problems in the short term. With a population of a few tens of millions, we could each have a big 4×4 or even a helicopter and still live comfortably and safely within ecosystem limits. Although I am not certain even of this, because this type of technically advanced consumption probably needs a specific 'pyramidal' production structure in state of permanent dynamic imbalance, with many small participants who have a dream of buying a 4×4 which is inaccessible to them.

Our economic and industrial system, like James Dean at the wheel of his car in *Rebel Without a Cause*, prompts us to push down hard on the accelerator pedal in the hope that we will be able to invent the wings that will allow us to fly off the cliff. But we have seen that such wings (technical innovations) will not have the capability or lift needed to avoid a crash. Like it or not, there only remains the very rational option to apply the brakes: reduce, as quickly and as drastically as possible, the average consumption of resources per person. It is not a question of knowing if we will return to the Middle Ages, but of knowing how one can return to a reasonable per capita consumption (not that of Middle Ages, because we have made some real technical progress since then, so keeping dental care for everyone!). The choice is not between growth and degrowth, but between imposed degrowth – because the resource issue will catch up with us in due course – or elective degrowth.

Since the high-tech system is driving us into the wall (or over the cliff, let's not quibble), why not try something else, taking the opposite approach and turning to low tech, to simpler technologies? Concretely, what might this mean? It is not very helpful to try to make a list of 'good' versus 'bad' technologies. The Amish have been trying for a long time, with varying degrees of success depending on the group, and through brain-numbing debates that take up long, candlelit winter evenings. As for high tech, its definition remains blurred.

When we speak about low tech, at this stage in our discussions it is paths to be taken and general principles that we must discuss, based on the conscious rejection of the hope of a successful outcome based on technological breakthroughs still to come. These general principles aim to effectively reduce our rate of consumption of resources.

Challenge the needs

The first principle has nothing to do with technology. Any human activity, other than hunting and gathering (and again, provided there are not too many people in a given territory), has an environmental impact. There is no such thing as a 'clean' car, or (totally) renewable, 'green' energy, or a 'zero-carbon' product or 'zero-emission' transport. Recycling is very important, but it is not enough and it cannot 'absolve' us of our material consumption. Energy will be used to recycle, material will be lost. It is important that we accept this and resist the *greenwashing* that we are hit with all the time, using our guilt to sell us more products, even trying to convince us that we are making a gesture that is beneficial for the planet.

A simple axiom: there is no product or service more ecologically sound, resource efficient or recyclable than the one that we do not use. The first question should not be "how to fill this or that need (or desire …)

in a more ecological way?", but "could one live as well, under certain conditions, without this need?" Then, of course, if the need cannot be avoided in this way without intolerable consequences, we must seek to meet it with the least possible resources. So, it is good to go first to the source, the root, of the savings, and in some way acquire the reflex of an 'ecology of demand' (reducing demand), rather than an 'ecology of supply' (green growth). The supply-side ecologist will call for the replacement of conventional power plants with renewable energies. The demand-side ecologist will propose to disconnect televisions first. The supply-side ecologist will request recyclable plastic coffee cups, the demand-side ecologist will have a cup in the office drawer.

In the animated film *Madagascar*, a New York lion and his zoo comrades, city dwellers to the ends of their claws, are welcomed into the Malagasy forest by a very cheerful king of the lemurs, who offers them the chance to return to the wild. "You mean 'live in a mud hut, wipe yourself with a leaf'-type wild?" exclaims the worried lion. To which the lemur responds, with a gleeful smile "Who wipes?"!

You should understand that I do not propose that we should go down this path. But why embark on a gigantic wind energy programme (which involves first concreting enormous foundations, opening access roads for maintenance, extracting and refining raw materials ...) with the aim of illuminating our shopping streets with Christmas lights on windy December nights? Why cover acres with solar panels to power energy-hungry billboards in streets or railway stations and flat screens in post offices? Why embark on complicated programmes of energy storage using hydrogen, polluting batteries, or pumped hydro in order to manage the intermittency of energy sources and thus keep our shop windows and signs illuminated at night?

I already hear supply-side ecologists (and those who do not even pretend to be ecologists) crying out about this: "This is ridiculous," they say, "we will not save enough electricity to shut down our nuclear power plants by switching off lights in shop windows at night! This is all out of proportion." Naturally, these are only a few examples intended to develop an argument. Apart from a few city centre retailers' associations, who could be against the removal of Christmas lights, or of flat screens in post offices? We didn't live so badly before we had flat screens in post offices, especially since they were installed just to reduce the perception of waiting: we always queued, but we are now less aware of it than before! In reality, there are many other possible 'sacrifices' that we can make, provided that we are ready to question some received ideas. The current mess is simply phenomenal, we can drastically lower our consumption without reducing our 'comfort' so significantly (see Part III).

The root of all evil

Beyond its undeniable effectiveness, the pure and simple suppression of need has another immense advantage: its simplicity. With a little political courage, we don't need to organize large industrial deployment plans, or to have thick reports produced, purely and simply to suppress advertising flyers! This principle of simplicity could also be usefully generalized by working at the source of the problems, rather than by trying to manage the consequences with standards and inspections, palliative measures and complex regulations.

Take the example of the recycling of glass. We have seen the phenomenon of 'degradation through use' in the case of metal recycling, with the mixture of small quantities of the minor non-ferrous metals in scrap of different sorts (see Part I). This phenomenon also exists for plastics and for glasses. For example, we obviously cannot remake clear glass from cullet (broken glass) obtained from the mixture of clear and coloured glasses, because the materials that tint glass (metals, like iron) are present in the mixture. Of course, it is not conceivable to re-sort the cullet 'by hand' after it has been collected, but expensive technologies, based on optical sensors, have emerged to do the sorting automatically by circulating the material on a conveyor belt. In short, is it better to throw all the bottles together, then have to rely on expensive ultra-technological facilities downstream, or sort upstream, 'at 'source', in different containers, clear glasses in one and coloured glasses in another, as is done in Germany? Or, even better, use only clear glass for all uses?

We can bemoan the low collection rate of single-use household batteries (about 46 per cent in 2018, 88,000 tonnes collected for 191,000 tonnes sold in Europe, and even lower rates at worldwide level). But if you wanted to combat the risk of seeing all these polluting disposable batteries reach incinerators, landfills or the wider environment, would it not be easier to ban them in favour of rechargeable batteries (for the sizes where they exist, which would exclude certain button cells)? Or do you sell new batteries to a customer only in exchange for the return of old ones, with some sort of deposit?

Similarly, to avoid a number of polluting and toxic products, it could suffice simply to decide not to produce them any more. The number of manufacturers of a given molecule is limited, economic stakeholders are often well known and easily identified, so a ban would be easy to devise and enforce. The downstream system would have to manage to adapt. Obviously, this would not avoid all pollutants, since some appear during manufacturing processes in factory discharges, incinerator particulates and so on.

In the case of climate change, the equation is quite simple: every tonne of carbon extracted from underground will in due course end up in the atmosphere. If we want to limit emissions, the only solution is to limit extraction (or, at the level of a country, imports), which can be done by quota or by tax, at the point of entry to the system (much more easily than at the exit), by incentives, restrictions, compensations, and various, varied and complicated mechanisms of exchange.

First liberticidal suggestions

Since I am now placed in the category of wet-blanket, killjoy, horribly liberticidal ecologists, opposed to the freedom for someone to stuff my letterbox with flyers pushing barbeque sausages as the promotion of the weekend, here are some examples of what we could implement, relatively quickly, by reducing our non-negotiable standard of living by just a little. Of course, this makes no assumptions at this stage about the social acceptability of the proposals, and how to make such choices in a democratic way!

First, some sacrifices could go almost unnoticed, such as a ban on advertising materials (in France around a million tons unsolicited in mailboxes every year, wow!), trainers that flash when you walk, or single-use plastic bags (many countries have already done this). Of course, the economic effects (resources consumed and pollution generated) are limited. Gains are modest, but they are so quick and easy to achieve, 'quick wins' as called in corporate language, why not go for it?

If we look carefully, we should even be able to find some very interesting and almost painless actions for ourselves as consumers. Take car tyres (rest assured, hardliners among you, I will tackle the car itself a little later – everything in its own time). Most light vehicle tyres are regularly exchanged for new ones (typically after 30,000–40,000 km), but a technique – called remoulding – exists to reuse tyre carcasses and to renew only the tread. It is used systematically for truck tyres, which can have up to a million kilometres of use after several remoulds. Although it is perfectly safe, the technique is used very little for private cars, much to the delight of tyre manufacturers. In France we throw away 250,000 tons per year, burned or put into landfill. If we adopted remoulding generally, if necessary adapting the design a bit and going so far as to sell new vehicles with remoulded tyres, we could save at least 160,000 tons of worn tyres a year based on a very cautious assumption of two remoulds on the life of the tyre, that would reduce consumption by two-thirds, with no difference for motorists.

Let's raise our level of criticism a notch. Quite frankly, bottled water, really? We cross Europe, and sometimes even the oceans, with water in our hands. Fukushima even made the Japanese drink temporarily some European water. Furthermore, we must suffer continuous announcements that the bottles are more and more frequently transported by train, are ever lighter, are ever more recycled into wonderful sweaters or garden chairs, in short, are more and more virtuous? At this rate, everyone will be submerged under a mountain of garden chairs by 2100! And what about the societal utility of advertising spending as a whole? Even more than flyers, the consumption of resources by advertising is very real: paper and ink in all those newspapers and magazines, advertising hoardings to renew regularly, and now flat screens, not to mention all the 'event' shows and billions of goodies, all these little advertising items ...

There are other potential 'mid-range' topics that we might consider. We could agree to limit ourselves to natural colours in newspapers, periodicals and brochures, clothes, shoes, furniture, packaging, toys and so on (according to our choice), to avoid dyes based on rare metals or synthetic chemicals. This implies shades of grey, ochre, beige, chestnut and unbleached colours. Newspapers could be printed again in black & white – we would reserve bright colours – blues, reds, oranges, greens – for particular and lasting uses only (clothes, bound books, household decorations, works of art ...). We could give up excessive use of machines on an individual scale. It is not a matter of giving up washing machines, but it might involve, for example, sharing electric drills and other tools in collective housing, while the electric juicers and yoghurt makers would disappear. In general, the aim would be to achieve a better balance between the use of labour and machinery. We might even consider the use of animals (sheep or rabbits to keep lawn grass short rather than lawnmowers: such a service already exists for companies and municipalities). I am always provoked when I meet a municipal employee using a noisy leaf-blower instead of a broom to clear up dead leaves in the streets ...

Let's raise our thinking up a notch. But be warned it could become really, really painful, although with the prize of huge savings in return. We could limit the speed of all cars to 120 km/h (75 mile/h) or even 90 km/h (56 mile/h). The reduced speed would of course lead to lower fuel consumption, but would also allow the weight of the vehicle to be reduced (because with lower speeds, there is less energy to absorb in the case of an accident, so less steel is needed in the car body), in turn allowing a further saving in consumption. Improvement can easily be 20 to 30 per cent, almost at the snap of a finger. Compare this to the painful evolution of current standards, year after year. For the more enthusiastic drivers among us, I'm not sure that you would lose so much sensation of speed, if the size of the car drops proportionally (think of kart racing!).

We could also manage cold weather differently, first and foremost by reducing the temperature to which we heat our buildings and, on the other hand, by not cooling them in summer. To remain politically correct, we always talk about reducing heating costs through better insulation. But we have seen the problems of the rebound effect, the number of houses that we have and the difficulties posed by the design of many old buildings. It would be so much simpler and faster to reduce the temperature, for example by limiting it to 16 or 18°C depending on the time and place. It is also much easier and cheaper to insulate and cover the body than to insulate and heat an entire room![2] Today we have clothes and materials that are particularly effective protection from the cold, certainly compared to some of what was available to our great-grandparents. Of course, this would imply some sacrifices. If you decrease the temperature in the office, employees will have to stop wearing light clothes when it is freezing outside. Some shared spaces (meeting rooms, bars and taverns, cinemas, libraries, museums …) could be heated a little better than private spaces. From time to time, we could collect in the shared places instead of everyone staying at home. To wrap ourselves up, thanks to the Swedes we will still have duvets that insulate well. The implementation of all this would probably be quite simple – by regulation in public and commercial buildings, and by a very highly progressive energy tariff in private housing.

Of course, under such a regime, we would seriously celebrate the return of spring! It will of course not always be pleasant, but what an enormous saving of energy! In France, 25 per cent of final energy use is spent on residential and business heating (40 out of 160 Mtoe – millions of tonnes of oil equivalent – total excluding non-energy use), and as energy losses increase linearly with the difference in temperature between inside and outside, lowering the thermostat a few degrees rapidly decreases energy expenditure (7–10 per cent for a single degree, on average).

Let us quote the thoughts of Ishmael, the narrator of *Moby-Dick*:[3]

> truly to enjoy bodily warmth, some small part of you must be cold, for there is no quality in this world that is not what it is merely by contrast. Nothing exists in itself. If you flatter yourself that you are all over comfortable [...] then you cannot be said to be comfortable any more. But if [...] the tip of your nose or the crown of your head be slightly chilled, why then, indeed, in the general consciousness you feel delightfully and unmistakably warm. For this reason, a sleeping apartment

should never be furnished with a fire [...] then there you lie
like the one warm spark in the heart of an arctic crystal.

Are you tempted?

In short, there is no limit to the number of ideas for ways to reduce our material consumption, from simple and consensual to more iconoclastic (Figure 2.1). Without going to too great an extreme, we could print newspapers and periodicals on a sufficiently 'soft' paper that it could then serve as toilet paper. The numbers come out pretty well: in Europe the consumption of newsprint is fairly similar to the consumption of toilet paper – why not kill two birds with one stone? A return to our roots in a way, since tearing of newspapers into sheets of a suitable size was a national sport between the two wars. Or sales of newspapers could be restricted to revitalized communal reading places, such as cafes or libraries, with a higher price per copy to ensure economic viability?

Figure 2.1: The 'ecolo-liberticidal' matrix

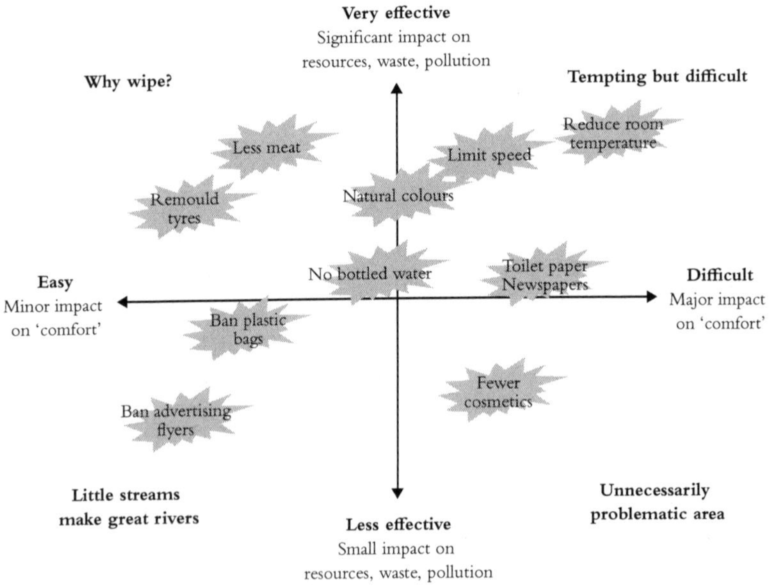

Note: Purely illustrative.

Truly sustainable design and production

Let's move on to the seriously technical issues. How can we achieve sustainable production? If we want to avoid future generations suffering an unbearable environment or a return to the Iron Age (and let's be optimistic, it is the Iron Age and not the Stone Age, because the Earth's crust is composed of 5 per cent iron, and 8 per cent aluminium!), we must drastically reduce pollution and 'net' consumption – the quantity we extract each year – of non-renewable raw materials. Since recycling has its limits (at each 'cycle' of consumption, some resources are lost and waste is generated), salvation therefore requires a considerable increase in the lifetime of products.

It is therefore necessary that products are designed and manufactured to be, so far as possible, resource efficient (especially concerning the rarest resources), non-polluting, durable, robust, easily repairable or reusable, modular, and easier to recycle at end of life. This represents an about-turn compared with technical or cultural programmed obsolescence, marketing differentiation and the logic that all is disposable.

To achieve this we need to address many aspects of products: the materials themselves, especially concerning the use of additives, complex alloys and composites; the design of objects and especially their modularity, the possibility of changing parts or reusing them at the end of life – including functional modules or even simple, standard parts such as screws; the 'reparability' of the products, the ease with which they can actually be maintained locally, by the user or the owner, or by a network of craftsmen, which implies, and we will return to this, no longer devaluing manual trades; and finally the regional scale at which the products will be manufactured.

Relocation, local 'reparability' and usability

In general terms, some relocation of production is necessary and even highly desirable, both for obvious energy reasons, by reducing the need for transport, and for social and environmental reasons, because it forces us to learn to manage locally the well-known 'negative externalities' of production. When you are a boss or a shareholder of a factory, it is much more difficult to pay slave wages or create shamefully polluting discharges if your customers live a stone's throw away and are themselves your employees, or relatives of your employees. Even on a small scale, reputation is a strong regulatory parameter. Have you ever boycotted a shop because the owner was treating his employees badly? A final motivation is societal, because local production will give meaning to work and allow a society to be constructed at the scale of the village, the city and the region ... where everyone finds his or her place in the economic life of the area, and is recognized – liked or disliked – by other members of the community.

Naturally, certain production and activities will need to remain non-local, even if only quarrying and mining or certain agricultural production all of which are highly dependent on geology or climate (nobody talks about making olive oil in Belgium). It is therefore not a question of stopping trade, but rather of focusing on what cannot readily be produced locally. This means approaching trade outside a region as was done in past centuries, but considerably facilitated by technological progress as regards modes of transport.

It all depends also on what one can call local. Even the simplest, most user-friendly, most repairable objects cannot be made entirely locally. Take the example of a bicycle. Even a basic model contains several hundred parts, most of which have a technical content that is not 'locally' manageable: metallurgy of alloys and dissimilar metals, machining and fitting of parts, vulcanization of rubber tyres, preparation of anti-corrosion paints or of grease for the chain ... ask a village blacksmith to make you a derailleur! However, once manufactured, it is clearly possible for ordinary people to fully understand its operation and to 'tinker' with repairs. A network of repairers with access to simple spare parts can keep it in shape for very many years, if not practically indefinitely.

We cannot say this, in their current design, about even the simplest of today's cars (with several tens of thousands of component parts), mobile phones or computers (with integrated electronics). On the other hand, a washing machine or a refrigerator could, under certain circumstances (simplification of the electronics, robust motor ...) fall relatively easily into this category, which is quite reassuring.

If we think about it, many everyday objects have become high tech and unrepairable, or even incomprehensible. Take the case of consumer medical thermometers. Remember the very simple models, the ones that disappeared because they contained mercury or later alcohol. They were impossible for a glassmaker to make 'locally' (put the liquid in the glass tube, calibrate it ...). But once manufactured by an industrialized process (and not necessarily a very high-tech one), they were perfectly simple and robust to use, and lasted for a very long time, provided they didn't fall and break. When they failed to work, we could understand why. Their basic materials were simple: glass, mercury or alcohol, some pigments for graduations. Their electronic replacements do not share these user-friendly features: they will eventually fail (display, battery, electronic circuit ...) and their content of raw materials, combined with their small size and the impossibility of a dedicated end-of-life collection arrangement, makes them impossible to recycle so they will eventually end up in landfill or being incinerated.

Good and bad standardization

To facilitate repairs and maximize opportunities for reuse, some standardization may be necessary. But there are certainly both good and bad examples of standardization.

Let's have an example of 'good' standardization. We could quite easily decide to only make three or four unique patterns/sizes of bottle (in clear glass, of course!) that could be used for all drinks and other edible liquids: water, milk, oil, wine, beer, sodas, fruit juices or even non-food uses (although perhaps, to avoid problems, we should dedicate some additional specific patterns to these). If there was then a general deposit/ return scheme, and local cleaning/bottling centres set up, then reuse of the bottles would be very attractive, both economically and ecologically. Under the current conditions, deposit/return schemes can be difficult to implement. Some calculations have shown that the transportation of empty bottles can consume more energy than is saved by their reuse, and I am quite willing to believe this when a bottle has to travel long distances to reach its original factory. But with standard bottles, they would be sent to the nearest plant, drastically reducing transportation needs.

In this way a bottle of Bordeaux wine, consumed in the Loire valley in France, might go to Normandy to be filled with milk, to be consumed in Brittany, and there filled with cider, and so on. Such schemes would be applicable in every country and region. From time to time there would be a little breakage, which would be collected in containers such as we use today, except that they would fill up much more slowly. Of course, without being able to use bespoke containers, none of this would be very good for marketing, the only differentiation being the label – in black and white or natural colours, please (see earlier in this chapter) – and the quality of the content. It would mean the end of the heavy whiskey or champagne bottles in distinctive shapes; and we would have to get used to the natural colour of red wine or olive oil (although some is already sold in clear bottles today). By the way, this is exactly how the system works today, for instance for Bordeaux wine: the only difference is the label and the rather large price variations!

For an example of poor standardization, consider the announcement made to a fanfare a few years ago (at least in official circles) by the association of mobile phone manufacturers. Taking a fervent 'sustainable development' approach they agreed to standardize … phone chargers. In addition to the somewhat ridiculous aspects of the case (a charger must contain some copper windings, but it is obvious that most of the pollution, upstream or downstream of manufacturing, is in the phone itself), my heart sinks when I think of all the entropy generated, all of the effort made – world experts and international administrative staff coming together in seminars after long business-class trips, thick reports, various commissions and agency files, conference calls – all for a mediocre result, which was anyhow swept away by the arrival of smartphones.

Poor standardization, therefore, is that which will probably take too long to arrive at and be too complex to implement, for limited effect, and that may even risk accelerating technical obsolescence through a change of standards. This is probably the case for construction, electronics, mechanics, where many components are anyway already standardized. For these sectors, the challenge lies in the reuse of components and modules rather than the standardization of new products.

Disposable and consumable products

Nevertheless, not all products that we use in everyday life are inherently durable. This is certainly true in the case of hygiene products, detergents and cosmetics, that are fundamentally dispersive, and packaging for certain uses (food, medical, hospital ...) that may be soiled or contaminated. It is therefore necessary to maintain the ability to manufacture some disposable and consumable products.

For disposable objects, a truly sustainable solution would be to manufacture entirely using renewable resources. Rest assured that I do not propose that we should use banana leaves for food packaging, although this was a solution that was adopted in tropical Africa before the arrival of the plastic bag. First, provided you have access to small shops or market-stalls (and have more time ...) needs can be greatly reduced through reuse of packaging for a good deal of shopping. Bring back empty egg cartons and glass yoghurt jars to the grocer to be reused, bring your own plastic boxes for prepared meals at the delicatessen or food stall, reuse paper bags for loose fruit and vegetables, buying things like pulses and honey in bulk (this has already begun in some chain stores). Pills could be delivered by pharmacists in reusable pill boxes, a common practice in many countries. Disposable packaging would thus be limited to very specific uses (for example, sensitive products such as vaccines). We could afford to use a (very) little bit of 'bio plastics', preferably renewable or maybe not (if it is a small amount ...) but certainly easily biodegradable, or easy to incinerate without emission of pollutants, if this is possible ...

As for consumable products, it would be quite simple to manufacture many of them locally from simple and natural ingredients. It would not be a case of making your own toothpaste (from ash, clay and mint from the garden), but rather a network of 'local apothecaries' could very well do it, and might at the same time address the issues of containers – we could visit them for soap, and to fill our pots with moisturizer and toothpaste.

This is also the only way to avoid the very diverse range of often harmful chemical compounds that are present in all these cosmetics. It should be said that any cream is an emulsion of oil and water, like mayonnaise. In order to maintain it in condition it is necessary to add surfactants such as sodium lauryl sulphate or sodium laureth ether sulphate (otherwise the mayonnaise separates), and preservatives to prevent the mayonnaise turning. These are antibacterials and antifungals – like the potentially carcinogenic parabens preservatives, which are starting to be replaced in some products by methylisothiazolinone (I am sure you feel safer now).

There are difficult physical constraints to overcome in cosmetics and therefore the composition of even 'organic' cosmetics, for which regulations are a little fuzzy, can make you shudder. We can't really use pure olive oil to moisturize the skin in the traditional way (it is very effective but its rancid and persistent odour would require big cultural changes and courageous users if they wanted to take the subway), but its preparation at home or at the local apothecary, then storage in a refrigerator to avoid the development of bacteria during the few weeks of use, would be a good alternative, allowing us to ban chemicals that are harmful to the body and to the environment after dispersion.

Orient knowledge towards the economic use of resources

Before addressing some other crucial principles, I would like to respond to an objection that I know will not fail to be raised, which is that, in contrast to high technologies and the 'knowledge economy', low technologies will necessarily be outdated, neo-Luddite, obscurantist, and in opposition to (scientific) progress. To do less (first principle) and in more sustainable ways (second principle) will necessitate turning our backs on innovation, knowledge and research.

In reality, it is quite the opposite, it will require knowledge and research, but oriented to different objectives than today's. Take organic agriculture, agro-ecology or permaculture, three similar approaches that can fairly readily meet low-tech criteria. They need few or no inputs (so they do the same or even better, in terms of nutritional qualities or yields, with less), create no pollution, have respect for or restore natural heritage such as soils, allow local control. These require, to be effective, solid theoretical bases in agronomy and microbiology, a fine knowledge of ecological cycles and of interactions between microfauna, fauna and plants. They require local adaptations of the methods to the different natures of the soils, to the variations in exposure of the land and in climates, to the varieties of cultivated plants.

If we are not to lose too much in terms of crop yields, the development of these techniques cannot be done without a development of knowledge and a sharing of experiences, of theoretical and practical learning, and of additional research which we will still have to accumulate. Much progress remains to be made in the future.

Another perhaps less obvious example, is that of the 'chemistry of life', necessary if we want to be able to use, more than we do today but still in a reasonable way, resources or waste from agriculture or livestock in place of oil and gas to make raw materials for manufactured products. This chemistry of life is much more complex because it needs to take into account the reactions and interactions of many different molecules, whereas current organic chemistry is essentially 'mono-molecule' (divided, after the refining of oil, coal or natural gas, into classes – C2, C3, C4, C5 – corresponding to the length of the carbon chain …) and 'single process' (reactions or series of reactions, with or without catalysts). In a way, we need to revisit what the ancients did intuitively and with rather poor yields, large losses of material and significant pollution in the industries of the old regime, such as tallow foundries, mineral or vegetable dyes, leather industry and tanneries.… But this time, thanks to the theoretical knowledge we have acquired since then, we could probably do it with much better yields, more control over processes and precision in output products, and a more optimal use of resources. But let us be clear, I am not throwing myself into the arms of the white biotechnologies of which we spoke at the end of Part I, because they seek a process perfection closer to that of classical chemistry, which could be different from a more complex approach described here.

Finally, there are areas where almost everything remains to be rethought and redone: a less absurd treatment of waste than we have today, recycling, eco-design …

Of course, there is a major issue concerning how the necessary research might be funded, because it will be difficult to file patents in most cases, and this is doubtless not desirable if one wishes to deploy useful and beneficial techniques as quickly as possible. Looking at it, I see no other option but to fund such research using public funds. Furthermore, we can probably afford this, because, in case we forget, in 'public–private partnerships', private actors are often content to bear the costs of the programme, that is to say essentially the variable costs, while the public – you and me, my dear fellow citizens – pay most of the fixed costs. Basically, in research, tax finances the car, and the private sector, which finances the fuel, decides, for proportionally a rather modest outlay, on the destination.

Finally, knowledge could be better shared than it is today, which is a factor of our democratic health. Production, instead of being hidden and concentrated, would be more visible, more locally controlled, by a larger number of people. Every consumer should understand the impact, the pros and cons, of his or her purchases. In agriculture and animal husbandry today, a very small number of experts, seed companies and breeders control the theoretical basis and the practical production of seeds and breeds, while tomorrow, every farmer would be a member of a network of seed and animal exchanges. To reduce the volume of waste and to ensure the return of nutrients to the soil, it will be necessary for everyone at home to be aware of and understand better the principles of the great natural cycles and practical methods of composting. Is this being too optimistic about human nature?

Finding the balance between performance and conviviality

Probably the main problem of the high-tech world that we are being promised is that, under the pretext of seeking ever greater effectiveness and technical efficiency, more complicated technologies are being developed that are often hungry for scarce and non-renewable resources, particularly metal resources. Furthermore, these high-performance systems generally require closer manufacturing tolerances (which will in turn require additional technological capabilities on the production lines), with consequential greater fragility and risk of breakdown, and more regular and specialized maintenance needs. The search for performance at all costs does not create a very resilient industrial system that would be resistant to the disturbances that will undoubtedly happen, for example geopolitical risks, supply disruptions, peak resources, political instability ...

If we really want to save resources, and make the system more stable under the uncertainties of the future, it will be necessary, one way or another, to compromise between performance and 'conviviality'. We need to extend Ivan Illich's reflections,[4] to build 'convivial societies' in which modern tools would serve the individual as part of the community, rather than serving a body of specialists – convivial societies in which people would control the means of production. Such an approach contrasts with a technological and systemic complexity that ends up being counterproductive (owing to prohibitive costs of maintenance, the fragility of 'techno-dependence', of negative consequences and costs being imposed on the rest of society) and turning against its users.

It would be better for some applications, surely, to lose some efficiency, but to be simple and robust, to use proven materials and technologies, and to increase the local capability to look after, repair, maintain and manage objects, tools and technical systems. It is, in a way, the situation in which former European colonies found themselves when they could not maintain state-of-the-art infrastructure or machinery, paid for by development aid (often with a few second thoughts on the implied exploitation of local raw materials), for lack of spare parts and tools or lack of local skills.

Therefore, before making massive deployments of renewable energy should we not think a little about the choice of our wind turbines? Might it be better to have 'village' wind turbines, less powerful and with a reduced range of operation, perhaps connected only to a local grid, but robust, technically basic and easy to repair, constructed from less exotic materials (I remind you that we still do not know how to recycle the materials of wind turbines that we are currently installing and which have a lifespan of less than 30 years)? Or these 5 or 7 MW offshore monsters, packed with advanced technologies, and their associated subsystems?

Specialists will tell me that it would be foolish to install smaller wind turbines that would require much more concrete and steel per kilowatt-hour produced. But perhaps everything is a matter of where we place our system boundaries: what if we include the entire system needed to operate these large high-tech wind turbines – the smart grids, logistics bases, paved access roads, connection cables – and the rare earth and copper mining and processing with the environmental damage caused? Would the advantage still be with this macrosystem? I wish we had asked the question, even if the calculation is far from straightforward. And, above all, they contribute to the myth of the availability of unlimited energy. Come wind or calm, rain or shine, we will install everything necessary for the lights to come on at the flick of a switch. Even if we have to extend the macrosystem to the Sahara or beyond.

In the design of a high-rise building, it will be better to avoid buildings that are too tall. Although height is good from the point of view of land use, consumption of energy and of high-tech metal resources is excessive. The greenest cars should definitely be inspired by the old Citroen model 2CV: low maximum speed, low weight, simple steel construction, no accessories, low consumption.[5] OK, they were noisy and engine efficiency left something to be desired, and I imagine that they would not pass current emissions standards – a large number of old 2CV models in our cities and on our roads would be a disaster for public health. But let's not throw the baby out with the bathwater and try to return to the Middle Ages but with dentists! We can take inspiration from the 2CV (from its low maximum speed, its lightness, its robustness, its simplicity), but incorporate improvements in emissions controls, engine efficiency and aerodynamics, although this is probably not so simple technically. But, as we have seen, the ideal remains to reduce both the number of cars and distances travelled.

This principle of accepting lower performance, ageing, reductions in efficiency and alternatives to replacement by new models at all costs should also apply to buildings, infrastructure and existing industrial facilities. This is one of the ways to combat rampant urban sprawl on greenfield sites. If you hear that to improve or implement such and such an activity, to make it more 'green' or 'sustainable' we must start by concreting over fields, take out your revolver. A new regional distribution centre will tarmac over a few acres of fields, but it is all towards the 'good cause' of shifting traffic patterns in the internet economy. Or new facilities at some ports, at the expense of some wetlands, to accommodate new maritime logistics necessary for offshore wind. Or new high-voltage lines to support the future development of renewable energies. The path to green hell is paved with good intentions. We must 'invest for the future', 'get over the tipping point' and it will certainly be better afterwards.... Is not it better to do with what we already have, although it may be less optimal, with the renovation of existing buildings and urban land, abandoned factories and car parks for deserted supermarkets?

Perfection is sometimes beautiful, in the arts, in music, in dance! But aside from these moments of pure magic – and if one ignores the gruelling exercises needed to train the joints of the pianist or the ballet dancer – life is mostly made of imperfections. We should take inspiration from this for our industrial system and our daily life.

Relocation without losing (good) economies of scale

It is both appealing and necessary to try to relocate some production to bring it closer to the places of consumption. In this age of the 'mad dance of the shrimps', where the same products, from steel to tomatoes, meet each other as they cross the Alps in opposing directions, we must be able to envisage this.

This raises the question of the scale at which relocation could or should take place. Should we repatriate Chinese production to Europe, from Spain to Estonia, from Ireland to Romania? Or promote the re-emergence of domestic industries – ah, the good old days of shop stewards and paternalistic factory bosses? Or go to a regional or even local scale with installation of small workshops and micro-enterprises in each municipality or village. There is of course no single answer, it all depends on what we will still have to produce, but we can make some assumptions. In order to do that we need to distinguish between process industries, manufacturers and network industries.

Process industries

Historically, industries, especially the process industries such as iron and steel and non-ferrous metal production, chemical plants and oil refineries, glass and building materials production (plaster, tiles, cement and so on) have always had a tendency to concentrate, going to ever larger production sites and serving ever larger areas.

This concentration made it possible to improve economic performance through economies of scale (reduction in the amount of labour and energy per unit produced, grouping of key skills at a limited number of sites and so on), but is also the result of a natural tendency towards the creation of oligopolies or local monopolies: by buying out the local competitor, we can raise prices a little and improve our profits. The tendency towards consolidation of sites is therefore limited only by the difficulties of transporting materials which are often heavy.

Thus, when products are fairly basic and have a low price per ton, transport is not likely to exceed a few tens or hundreds of kilometres, except when there are possibilities of access to sea or river transport, which allow the transportation of huge quantities at very low cost. Hence the market for cement is based on local quasi-monopolies in the interior of a country, but around ports imported cement can be transported at reasonable cost. If the products have a higher price per ton, such as specialist steels or glasses, they can justify long-distance transport very well, easily exceeding 1,000 kilometres.

Some of these industries are already, by their very nature, relatively close to their consumer base. For others, it is probably not sensible to want to 'decentralize', for at least two reasons. First, by reducing the scale of production, we risk losing on energy consumption. However, that risk is not definite: the chemical industry is already envisaging smaller, more local units based on 'microreactors' and 'process intensification'. Nevertheless, it is likely that smaller units will require more raw materials per unit of finished product, primarily for the equipment. Second, and importantly, because we would create new 'collateral' environmental damage in the creation of new local production facilities: we would have to find new brownfield sites, install new high-tech equipment and consume space and resources now, all for a future expected gain in transport costs.

It would be better, once again, to 'make do with what we have'. The idea could be to accompany the drastic reduction of our material needs (generated by our virtuous behaviour to come) by the closure of certain sites or, better, the reduction of the capacity of the various sites (closure of facilities at the site or intermittent operation, as the case may be).

Let us take as an example the use of steel, aluminium and plastics in the various economic sectors (see Figure 2.2). The automotive, construction and packaging sectors alone account for 60–65 per cent of total consumption. This is not surprising since cars and construction are two of the 'heavyweights' of many economies – when the construction sector goes well, so does the economy, while the automotive industry was the driving force of Western industrialization from the 1950s to the 1970s. The third heavyweight – supermarket chains – relies heavily on large shops and packaged products.

But imagine that, if we were ambitious, we could reduce the need for resources by 50 per cent in construction and 80 per cent in automobiles and packaging. Without going into detail, the 50 per cent on construction could be achieved, for example, by doing without new offices and industrial buildings (make do with what we have), and with a little more rehabilitation and better use of the existing housing stock (in Europe there are 11 million houses unoccupied).[6] The 80 per cent reduction in the automobile industry could be achieved by limiting the total number of vehicles and increasing their longevity. In packaging, 80 per cent could be achieved by widespread bulk sales and the use of reusable containers.

The exact figures are not important, it is only a question of giving an order of magnitude. In these three sectors alone, we could save 40–50 per cent of the total consumption of steel, aluminium, plastic and cement.

Figure 2.2: Consumption by 'heavyweight' sectors

Percentage of resource consumption by sector

	Automobile	Construction	Packaging	Total
Steel	~20%	~40%	low	~60%
Aluminium	~20%	~20%	~20%	~60%
All plastics	~10%	~25%	~30%	~65%

With the financial crisis, it is what has already happened a little, hence the closures of some blast furnaces around Europe that followed. As car production declines (not by the reduction in their number but by the ageing of the fleet and the rise in imports), the need for pig iron has significantly reduced. In a world of low tech, economizing on pig iron and recycling scrap, we could probably make do with one or two blast furnaces to supply all of Western Europe. But don't ask me which ones to keep.

So far, we have discussed direct consumption in final products, and we should add to that what is necessary for their manufacture: robots, machines and assembly lines for cars, cranes, construction equipment and concrete shuttering for construction. This indirect part is relatively limited in mass consumption as in the motor industry, but not so for 'intermediate' activities. For example, the oil and gas sector, invisible to the end consumer, already absorbs 5 per cent of world steel production, for exploration platforms, well drills and casings, pipelines, oil tankers for transportation. This will be more challenging in the future, as we struggle to manage with less accessible resources: there is a difficult road ahead with longer pipelines and deeper offshore platforms, and/or with more wells to be drilled per barrel of shale oil and cubic metre of gas.

In summary, there may be some relocation of process industries but certainly not all of them. One cannot imagine installing a small blast furnace or a mini-cement plant in every average city or region of Europe. But why not, after all, for manufacturing plants whose scale can be reduced more easily, like soap factories or breweries?

Manufactured products

Even if a return of craft industries is eminently desirable, it is clear that a certain degree of mass production in the manufacturing industry allows enormous productivity gains, through specialization and division of labour, as Adam Smith showed in his famous example of the manufacture of pins. By dividing the manufacture of a pin into 18 separate operations[7] (draw the wire, straighten it, cut it, sharpen the point, strike the head ...), the rate of production is orders of magnitude greater than can be achieved by the most skilful workman carrying out all operations sequentially.

This effect of specialization should not be confused with the fundamentals of Fordism. For Adam Smith, the interest is in specializing a task to allow it to be carried out automatically and with dexterity, leaving the same worker to change task from one day to the next, or even in the course of the day, to work in batches on the repetitive tasks of the same production – a work organization that is not necessarily incompatible with the cognitive health of manual work. For Henry Ford, by contrast, the interest in the assembly line was to be able to operate with workers without special qualifications, who were consequently easily replaceable and who could be paid less. The aim of Fordism is thus not primarily to gain in productivity, but to gain from reduced wages of workers at constant productivity. The emergence of Fordism was, moreover, not easy: 'So great was labor's distaste for the new machine system that towards the close of 1913 every time the company wanted to add 100 men to its factory personnel, it was necessary to hire 963.'[8] There are therefore different types of specialization and different ways of achieving productivity.

Specialization, and the transport of semi-finished products which is its natural corollary, are thus not a recent development, as this 19th-century quote shows:

> What a cutler must strive to achieve is low costs. The penny knife [...] is the leading example of this kind of utensil. The blade is fixed by a rivet in a wooden handle. Despite its simplicity, it requires the participation of numerous factories, located in three different places. The handles are made in the Jura, the blades in the Dauphiné, while assembly is done in Saint-Etienne.[9]

These locations are some hundreds of kilometres apart.

Today the collapse of transport costs (owing to oil and containerization) has allowed production to be organized on a completely different scale, globalized and at a level of concentration and specialization never before imagined. It is hard to believe that the optimum is represented by the geographical concentration of, as an example, clusters of Zheijang province in China, in the Shanghai hinterland, with entire cities devoted to mono-production, like Datang, where a third of the world's socks, or Qiaotou, where 80 per cent of the world's buttons and zippers are made. And what about the capitalist concentration of the giant Foxconn (the famous subcontractor of Apple and others) that produces 40 per cent of all the world's electronics?

Around 60 per cent of watches sold worldwide, 55 per cent of cameras, 70 per cent of frames for spectacles, 80 per cent of duvets and umbrellas, 75 per cent of toys are made in China. I find it hard to believe that Adam Smith would feel at home there, any more than his almost contemporary David Ricardo and his (attractive on paper) theory of comparative advantages. The 3 billion Hangji toothbrushes are manufactured in several factories, the largest of which (Colgate) produces one third. Is it necessary to produce 1 billion brushes a year in a single plant to achieve the optimum productivity and economy? In reality, such production is only possible with many parallel (and high-tech) manufacturing lines, so I'll bet that smaller units – at equivalent slave labour costs – can achieve the same productivity. This is indeed the case for much of the other production of Chinese clusters – summarized in Figure 2.3 (although note that these figures are over 10 years old, dating from 2006 and 2009) – where cities and districts are specialized, but production (such as Datang socks), takes place in a multitude of medium-sized enterprises or small family workshops, based on a few machines.

How can such a system, such a concentration of activities, be sustainable in the long run? When the latest 'technology transfers' have been made (if we still have the slightest advantage now), what will we have to sell in exchange for all these manufactured goods? Tours of the Eiffel Tower, the Colosseum or Westminster Abbey, cancan shows or art galleries, the sale of luxury bags that are not even manufactured here? Producing olive oil where the sun shines is not to be criticized – provided that it is produced in a decent manner for the environment (which automatically disqualifies palm oil, displacing the last tropical forests) – but why, in the long run, should shoes be produced elsewhere than locally, except perhaps in countries where geographical constraints do not allow production of enough leather for the needs of the population?

It is therefore logical, desirable – and conceivable, we will see how later (see Part IV) – that a number of small to medium-sized factories will relocate to our lands. At least for those everyday objects (clothing, shoes, crockery, small tools, utensils ...) that do not require a priori too much investment, or too much machinery, because for high-tech objects such as consumer electronics, it may be more difficult to achieve critical size to maintain a high level of productivity in small units.

It will not mean a massive return to craft manufacture, even if such production will obviously develop. Better to keep a lock factory than to try to make all the necessary parts for their manufacture at the local blacksmith (who is already busy with your bike's derailleur ...). Better not to return immediately to the spinning wheel, as advocated by Gandhi. By 'de-mechanizing' to that extent we would lose a lot of production capacity – indeed too much – for many commodities, without necessarily gaining much on the consumption of energy or raw materials

If we want to take advantage of our reduced needs to progress towards a society in which we have more free time, it will be wise to keep those small factories and specialized workshops in which small scale does not prevent high productivity ... and to encourage the

Figure 2.3: Too far for Adam Smith? The Chinese clusters

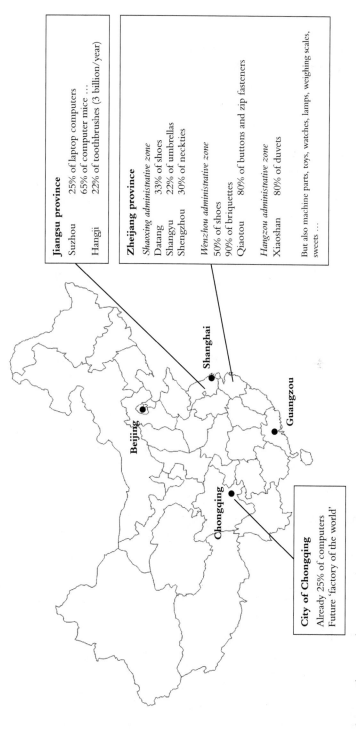

Jiangsu province

Suzhou	25% of laptop computers
	65% of computer mice …
Hangji	22% of toothbrushes (3 billion/year)

Zhejiang province

Shaoxing administrative zone
Datang	33% of shoes
Shangyu	22% of umbrellas
Shengzhou	30% of neckties

Wenzhou administrative zone
	50% of shoes
	90% of briquettes
Qiaotou	80% of buttons and zip fasteners

Hangzou administrative zone
| Xiaoshan | 80% of duvets |

But also machine parts, toys, watches, lamps, weighing scales, sweets …

City of Chongqing
Already 25% of computers
Future 'factory of the world'

Sources: Kham Voraphete, *Forces et fragilités de la Chine: Les incertitudes du grand Dragon [Strengths and weaknesses of China: The uncertainties of the great Dragon]*, L'Harmattan, 2009, Lu Shi and Bernard Ganne, 'Understanding the Zheijian industrial clusters: Questions and re-evaluations', International Workshop Asian Industrial Clusters, Nov. 2006, Lyon, France.

shoemaker and other craftsmen to concentrate on repair rather than custom design.

We still need to look at technical aspects of the manufacturing itself. We need to remember that labour productivity, while dependent on organization, is also essentially based on increased energy and material expenditure. Machines, robots, production lines, machine controllers – the technological content of factories is constantly increasing. In order to try to stay ahead in a very competitive world environment, they too are impacted at an increasing rate, by technical obsolescence. A key question is how to preserve manufacturing productivity while reducing energy consumption and technological content?

It is clear that, historically, great leaps in energy use were made through first using water mills and then steam engines in industry, shredding rags for paper-making, processing minerals, and driving looms. Some small industries have been energy-users for a long time, thus in the deepest France of the 1860s 'in the factories, shafts are driven by machines; but in light industry, in cottage manufacturing, they may be set in motion by dogs, idiots and blind men, using treadmills'.[10] Rest assured, I do not propose anything so drastic, except perhaps with dogs or indoor joggers.

Of course, as elsewhere, improvements in productivity through the use of energy and automated machines are probably subject to diminishing returns. The gains from switching to the first, relatively simple, machines were much larger in percentage terms than can now be achieved by reorganizing a production line or investing in a state-of-the-art system. So, there may be a need to accept some decline in labour productivity, but is this such a big problem in a society that no longer provides enough work for its entire labour force?

By developing small workshops and micro-enterprises, equipped with some simple and robust machines, but retaining some of the high technologies that have been acquired (numerical control of machines?), it should be possible to maintain a good part of the current productivity, while lowering the energy content. This local production system could be supported by a network of craftsmen who, in addition to providing the usual maintenance and repair of manufactured objects, would allow basic skills to be regained.

Network industries

There remains the thorny issue of what we might call network industries – the utilities but also various systems providers. Perhaps more so even than manufactured objects, these are an essential part of our economic activity and our modern comforts – distribution of water, gas and electricity, sewage networks, public transport, hospitals, telecommunications. This comfort can be so commonplace that we are not really aware of it anymore (yet, when we think about it, to see water running from a tap is magical). But we are quickly reminded if it is ever temporarily missing. Like happiness, it is painful when it suddenly disappears.

Going beyond our comfort, what is at stake in the maintenance of these systems and services is our capacity to live in very populous communities. We are now very numerous, much too urbanized, crowded together into limited spaces. At several hundred inhabitants per square kilometre (the European average is $106/km^2$, but this includes the mountainous regions of Sweden, the Pyrenees, parts of the Alps ...), we cannot think of collecting firewood from around our houses, taking water from rivers and streams for our laundry, or defecating on the side of the road. The physical concentration of people (around 600,000 passengers per day at Paris Gare du Nord station, the busiest in Europe) and the risks of microbial and viral contamination are such that to avoid epidemics we need to adopt strict hygiene rules (for our bodies as well as our places), maintain impeccable water supply and sanitation systems, and achieve a high level of health care and security systems. On top of that, our industrial system is creating its own dire problems: poverty and destitution, rainwater that picks up atmospheric pollutants so that it is no longer drinkable, new viruses from the concentrated breeding of chickens and pigs.

In the first place, we must question the dogma of competition as beneficial to service provision. In the network industries, it is pretty detrimental from an environmental point of view, but also in economic terms, whatever one may say. We have multiple sets of equipment and parallel installations (consider mobile telephony and its relay antennae, even if from time to time different operators share a tower ...), the unnecessary expense of duplicating marketing/sales/communications activities without changing anything at the customer end (a frightening example is in the 'liberalization' of electricity supplies, where the same power distribution company provides the service at one end, that is to say the only service seen by the customer, while the historic generators must sell electricity at wholesale prices at the other end! What's left in the middle besides marketing and billing services? A mystery). Competition therefore generates increased material consumption (even if only advertising media – but it is true that sometimes, even in a monopoly situation, it is good to keep the media quiet by buying some advertising space),[11] with no obvious gain except perhaps a stimulation of innovation (and even that is questionable if we consider videotex systems such as

Minitel in France or Prestel in the UK and other services that dated from the time of public monopolies).

For ideological reasons, we have deliberately confused the 'healthy' competition between three bakers in the same locality with this counterproductive competition between networks, leading to incredible and senseless cuts within systems: railways, distribution of parcels and mail, electricity supplies. As for competition by concession (renewal by call for tenders) in other cases such as water and sanitation, we also see limits and risks, related to local corruption.

Second, technical complexity in network industries is very high and pathways for development of our low technologies are far from obvious. Major high-performance equipment (alternators, compressors, pumps, motors, automation and so on) is widely used and electronics and computers are ubiquitous (surveillance networks, regulation, control ...). We cannot imagine all the technical elements that must be regularly replaced on both the vehicles and the track and signalling networks to run a simple train. The complexity is comparably high for telecommunications equipment, or that necessary for medical examinations or surgical procedures. Technological dependence continues to increase steadily, either because we are increasing our demands (for example, users have rapidly become dependent on the signs that announce waiting times for metros and buses, thus driving the rapid deployment of the technologies), or because, under the pressure of costs and the associated reduction in the workforce, we are mechanizing at a hectic pace (for example, the driverless metro, or services for remote maintenance) – discussed further later, see p 76. In addition, the networks are also dependent on a complete technical macrosystem (the manufacture and logistics of spare parts need industries and transportation systems, maintenance technicians need vehicles and transport systems ...), and are interdependent with other systems (telecommunications equipment needs a power supply, the power network is dependent on remote communications).

Some solutions can of course come from a reduction of needs at source. Treatment of waste can be reduced by composting and reduction of packaging, energy by lower consumption and transport by reduction in speed or by automatic provision of information. In villages and small towns, another part could come from empowering homes or neighbourhoods. Thus, alternative solutions exist to the sewer (dry toilets for the purists or filtration by plants in successive basins, although this requires a lot of space), or for energy (small-scale renewable generation – solar thermal, micro-hydro ...). These local solutions would reduce needs, or allow solutions with lower technological content to be implemented. But they are unthinkable at the scale of large urban agglomerations.

I admit that it's a little thin. However, the overall resource utilization of these network industries is likely to remain quite low compared to the large consumption sectors such as automobiles, construction and consumer goods. For example, the rail network in France is 31,000 km, about half a metre of track per person, 1 or 2 metres per household. Even taking the rails, the track bed, bridges and other structures, the overhead cables and the signalling all together, the metallic content (in weight as well as quality) is much greater in a house or in a single car (with a shorter life). But it may be otherwise if we add together all networks, including gas and water pipes, electrical cables and mobile phone masts, machines in hospitals and so on. This will be especially true if exponential development continues (as in telecommunications of course, with the explosion of internet transmission equipment, corporate servers, cloud computing, data storage and data centres …) or if we prepare for a massive deployment of renewable energies, with smart grids linked to 'intelligent' (and communicating) smart meters that would supposedly allow us to balance a necessarily intermittent supply at all times with variable consumer demand, by switching off your refrigerator when the wind drops. We are not there yet, but the good fairies of green growth are already leaning over the cradle.

Whatever happens, unless there is a sudden and complete disruption of society, for example through financial collapse, there is little chance that resource shortages will arise so suddenly that network services fail overnight. We therefore don't need to fear a brutal return to the water well and candle. Drinking water will not be the first thing to fail, even with expensive oil, for it is better to give up one's car than to have to fetch water from a well, and it is better to have to put on a pullover than to light oneself with a candle. If restrictions emerge, people (or the state) will naturally arbitrate toward those with the highest social acceptability. Firefighters will not run out of fuel, but the average quality of hospital care will deteriorate gradually. There will be many more things going on before we need to re-dig our wells.

But let's be careful not to increase the technological content of service networks, for example by avoiding excessive automation. But if their impact on resource consumption remains low – telecoms aside – we probably have a lot more to gain in the manufacturing sectors and from mass consumption. The efforts to reduce our service network needs would be disproportionate to the savings generated, and the needs are, in any case, often related to our strong urbanization, which is not a simple issue to address either.

De-mechanize services

Machines are useful for production, naturally, to increase productivity but also to combat the arduousness of work. But why replace humans with machines everywhere? In recent years we have also witnessed a proliferation of machines in services.

I am not talking about the fact that it is now necessary for many businesses to use computer hardware and networks. The travel agency and the design office use them systematically. There are probably some places where use is perhaps a little excessive, and where we could accept a reduction in the amount of technology that we use, such as those terminals used by waiters in restaurants, with which we imagine gaining a little productivity and avoiding errors in the bills. We could easily stop using those overnight. I admit, however, that it is difficult to imagine ordering a train ticket 'manually' in the near future, without a well-functioning computer network and software, especially if we want to retain the ability to reserve seats. Applied to such services, technology is now embedded and computerization has been developed since the 1970s at least.

On the other hand, we are witnessing more and more the gradual replacement of reception and service occupations by machines or automatic kiosks: in railway stations and subway stations, post offices, government agencies, ticket distributors in museums or zoos, or in supermarkets. In public places or establishments open to the public, beverage vending machines proliferate.

From the resource point of view, nothing is more detrimental. Simple human work is replaced by consumption of materials and energy. Machines and screens are filled with electronics, and thus permanently tie up rare metals. Of course, it may not be very rewarding work (for the ticket inspector and supermarket cashier, for sure, although it is probably less true for the village post office counter, which had a real social role). As we proceed, we replace the old low-skilled jobs by others. There is much to be said about the deterioration of these new 'non-robotizable' service jobs, this neo-proletariat responsible for delivery rounds to supply coffee and confectionery distributors in stations, to change businesses' water cooler bottles, and to maintain all these complicated, fragile and capricious machines – because without maintenance, they would all quickly break down.

Mechanization, technologization and robotization have their limits, because it is always necessary for somebody to physically carry out maintenance and to supply spare parts. It is even more necessary for a complex system, because there is also the question of who maintains the equipment of the maintainer, and so on. As long as we have not seen one robot troubleshoot another (we are a long way from that, I think), the world is likely to remain very human. This makes me doubt the potential risk of future hordes of 'self-replicating nanobots' invading the world without asking our opinion. But I may not be correct in this, although nanotechnologies have many other disadvantages that are enough to make us abandon them, as we have seen previously.

It is therefore impossible to do without human work, even in the 2.0 economy, because 'we cannot hammer in a nail with the internet'. Labour will always be needed for refuse collection or for building construction, since we can't 3D print all parts of houses or buildings. Conversely, thanks to the internet, service businesses are relocating more and more easily, to the Maghreb for French speakers or to India for Anglophones. Following low-skilled occupations such as call centres (where it is usually enough to follow a script, or instructions delivered via software), it is the turn of design offices and software developers, and in due course on the list will be 'high-value' trades, such as lawyers or radiologists. Oh yes, get ready, soon the radiology department in your hospital may send your X-ray to India where it will be interpreted at low cost, before the diagnosis comes back electronically to be signed and endorsed locally by a well-qualified European radiologist. And who will without doubt continue to make you pay a high price.

Until now, we have stuck to the historical pattern of the division between blue-collar and white-collar workers, between unskilled and skilled jobs. And, inevitably, in this pattern the key to escaping unemployment was to increase one's qualifications. But this arrangement is about to be swept away. The old split will be replaced by that of relocatable or non-relocatable employment.[12] Do you dream of the top engineering schools for your son or daughter? Then he or she should choose their specialty well, because China, India and Indonesia are training engineers in droves, and at rather lower cost. Meanwhile it is well known that a plumber working for himself or herself, on what is certainly a less prestigious job in society, earns a lot more per hour than many 'qualified' service jobs, as can be seen from the plumber's bill.

Let's go back to our automatic terminals. As well as being disappointing in terms of resources and energy, they are also aberrant from an economic point of view. Our 'decision-makers', convinced that the number of imperfect workers must be reduced as quickly as possible, especially in these often-unionized occupations, demonstrate frightful short-sightedness. What is happening in practice? In order to be more 'efficient', and starting from the good intention to try to serve customers better and at less cost, local employment is destroyed and the trade deficit worsens (because most of the physical content of the machines is manufactured abroad, and their energy is purchased externally), helping to accelerate the destruction of the foundations of our societies.

Is it possible to 'backtrack'? Yes, definitely, but we should be careful not to go too far (I'm not sure that we need to re-hire ticket-punchers). With millions unemployed, we can certainly re-humanize some services without changing anything at all across society. Let's keep some very useful machines, and not recreate the occupation of washerwoman to replace washing machines. Although one could argue that the wealthy of the developed world in effect subcontract the washing of their laundry to 'virtual washerwomen', in countries far away whose labour force makes our washing machines or extracts the non-renewable resources that are used to manufacture and operate them.

Knowing how to stay modest

As you will have noticed, these are only pathways, some scaffolding and incomplete reflections. I do not pretend to have an answer to everything, nor to what or how. That is good because it is in keeping with a final principle of low technology, that of knowing how to remain modest.

The discoveries of science and technical progress have been so dazzling that we have lost our bearings. Our 'Promethean' science today promises us nothing less than omnipotence: pushing the limits of life with the medicine of 'red' biotechnologies, choosing our descendants with genetic engineering, even seeking eternity through the 'transhumanist' movement trying to download our brains onto hard drives, or to perpetually clone us. We seek to transform the inert into living material, to create life from scratch with synthetic biology. We seek access to universal knowledge (but perhaps not to wisdom …) with databases, translation software, and the interconnection of all human beings.

Such promises of science and technology are not new. They have provided nourishment for popular science literature and science fiction since the 19th century and even before. What is possibly new is that science now promises to repair the environmental damage it has generated. We will 'heal' the planet, while conquering the universe. An ambitious programme!

Undeniably, the incredible progress made in many areas makes some possibilities quite believable, but only if we take into account all the limiting factors that have been presented previously. Whatever happens, all of this will be DIY in relation to the complexity of the world around us.

Some spiders, if interrupted as they weave their web, are unable to complete it or to resume work in progress on a web already begun. They can only build it from 'A to Z', complete in one pass. They are a kind of 'genetic automaton', with all their behaviours 'pre-wired' from their creation. This is true of many more 'evolved' animals than invertebrates, such as the salmon returning to the river where it spawned, or the common swift, whose parents begin their migration without waiting for it, and when it flies for the first time will next land two and a half years later, when it then nests itself – before almost immediately embarking on a journey of 6,000 to 8,000 kilometres to reach the south of Africa!

How are all these such complex behaviours recorded on DNA strands? Let's re-learn humility in science! Whatever the progress of molecular biology, and even if the famous Moore's Law in computer science continues, we will never be able to discover, model, or understand how a set of chemical bonds on macromolecules can encode the steps in the making of a spider web, which is the work of hundreds of millions of years of evolution. With recent advances in computation and data storage and the emergence of 'Big Data', masses of data are being trawled to establish statistical correlations (which may be very useful, of course), but not cause-and-effect relationships. We see, but we do not understand.

What is left for us if we are too Darwinian to throw ourselves into the arms of religion, which brings simple answers to the mysteries of the spider? Let us accept that we do not have answers to everything. Let us contemplate the passing swifts ... and reclaim the poetic and philosophical dimension of the world. If science cannot do it, only poetry and philosophy can help us to describe and understand the reality that surrounds us. These two subjects, which formed an essential part of the teaching of the ancients, are too sorely neglected today.

We cannot control everything, and that will also be true of a transition to a world of low technologies. The road is not clear, everything is now so confused, so 'systemic' with its positive and negative feedback loops, that it is not worth trying to unroll a plan developed in advance, since at best nothing will go exactly as planned. As a ship in fog on a rough sea with an imprecise compass, one can get an idea of the direction of dangerous coasts to avoid, and try to steer a course with some clear objectives. We can thus usefully remember our seven principles, shown in Figure 2.4. Seven is always a nice number because it is a little mystical in Western cultures, but we could surely find two or three more for fans of the Ten Commandments.

Figure 2.4: The seven commandments of low tech

1	Question the need	→	Thou shalt ask thyself, why wipe?
2	Design and manufacture for true sustainability	→	Thou shalt make simple and durable. Thou shalt remember that everything has an impact
3	Orient knowledge towards economy of resource use	→	Thou shalt seek and transmit knowledge in the right direction. Thou shalt also inspire thyself from ageless knowledge
4	Search for the balance between performance and conviviality	→	Thou shalt be satisfied with less beautiful or new, designing for lower performance
5	Re-localize without losing the good effects of scale	→	Thou shalt re-localize with finesse, at the right level
6	De-automate services	→	Thou shalt replace people by machines with caution
7	Know how to remain modest	→	Thou shalt marvel at the complexity of nature

PART III

Daily Life in the Era of Simple Technologies

It is now time to sketch out what the application of these worthy principles might do to our daily lives, without yet making any assumptions about their political, economic, social or cultural feasibility.

Of course, you shouldn't be too sceptical as you read this, thinking that these measures and structural changes are unreal, irresponsible and utopian. Of course, they are alarming, out of line with current trends and societal norms, damaging to jobs and clearly impossible to implement in the context of global competition. We will address these crucial issues in Part IV.

At this point it is also unnecessary to describe in any detail the conditions and mechanisms for transformation, to speculate on the role that might be played by various stakeholders (political, social or other), or to list the concrete steps to be taken, the laws to be enacted or the costs and benefits of particular changes. It is not yet the time for that. Let us for the moment set out what needs to be changed and the general directions that could or should be adopted. Let us see where we should go before we consider how to get there; let us try to develop some conviction. The political, fiscal, regulatory and diplomatic measures and the like, and their cultural and moral implications, will follow more easily, but we should not forget the lessons of the past, namely that the end does not necessarily justify the means.

For the moment, let us imagine, purely as a hypothesis, that we can find a way to reverse the current worrying trends, and let us immerse ourselves for a moment in everyday life in a time of simple technologies. How could and should we feed, move and house ourselves, communicate, trade and consume? In short, how might we live?

Agriculture and food

Let us start with food because, on the one hand, we have seen that feeding a growing world population is never going to be simple, while on the other, human nature is such that one approaches almost everything with a greater serenity if one has a full stomach. According to the old adage, only three meals separate a society from chaos.

The challenges

In the 'upstream' activities of food production (agriculture, livestock, fishing), we need to be able to feed humanity in decades to come without compromising our capacity to feed for the coming centuries. If possible, we need to do this by producing quality food that is beneficial to health, while reducing the environmental impact of agricultural activities, in particular avoiding devastation of the remaining natural areas (currently being transformed at high speed into arable land to compensate for soil depletion) and agricultural land (being reduced to biological deserts through the misuse of pesticides). It is therefore necessary to maintain or improve agricultural yields while at the same time reducing the need for inputs: of synthetic fertilizers (nitrates), mined minerals (phosphates and potash) and pesticides (herbicides, insecticides, fungicides ...). In the 'downstream' activities of the food processing and distribution sectors, we need to reduce the amount of transportation, and the packaging and waste that is generated.

Let us be clear, we have not really begun to do this, either nationally or globally. Between 1945 and 2007, the world consumption of pesticides increased from 0.05 to 2.5 million tonnes per annum (and it is expected to grow to 3.5 million tonnes by 2020), while over the same period the toxicity of the products used increased by a factor of 10 (that is, 10 times less pesticide is needed to obtain the same effect)! There are currently more than 900 active substances listed, with 15 to 20 more added each year, making any serious study of the effect of interactions between combinations of products impossible. Europe is world champion at 3.9 kg per hectare per year (kg/ha/year) while the world average is 1.5 kg/ha/year. France is European champion at 4.5 kg/ha/year. This is partly explained by a long tradition of viticulture – a quarter of the pesticides are spread on grape vines, a crop that is very sensitive to disease and to infestation.

Fertilizer consumption is also increasing globally (from 30 million to approaching 200 million tonnes between the early 1960s and today), as the area of cultivated land has grown, although the weight per hectare (110 kg/ha/year on average) has tended to level off or reduce as a result of a more rational and economical use arising from a better understanding of soil composition and dynamics.

Agricultural productivity and yields

Let us start by focusing on the fact that we often attribute increases in agricultural productivity to a little bit of everything, from the green revolution to the decline in the number of farmers to the benefits of mechanization. However, yields and productivity should not be confused (see Figure 3.1).

In agriculture, yield is production per hectare. This depends on the nature of the soils and climate and, first and foremost, the varieties that are cultivated. It can increase through selection of appropriate varieties and cultivation methods, and through the use of fertilizers to promote growth and pesticides to reduce losses.

Agricultural productivity is production per worker. This depends on the area a worker can manage, multiplied by the production per unit area, that is, the yield. At constant yield, productivity increases by moving from manual cultivation to use of draught animals, then mechanization, which allows an ever-greater area to be cultivated with fewer people.

When we talk about prices, production cost and competitiveness of our agriculture in a globalized system, it is productivity that will interest us. However, if we want to feed the planet, regardless of the number of hands we put to work, we need to talk about yield per unit area.

What is happening is that productivity is continuing to increase, both in 'developed' and in 'emerging' countries, through ever more mechanization (more powerful and sophisticated tractors, combine harvesters and other machines, GPS guidance and so on), larger field sizes, and fewer farms overall. But, for some crops, the yield per hectare has begun to stagnate, as the example of common wheat in France for the recent period shows (Figure 3.1). The 'miracle' of agriculture in recent years – at least in France – has simply been production of about the same amount with fewer staff. However, this has been at the cost of environmental (eutrophication, exhausted soils, tropical deforestation …) and societal problems (desertification of the countryside, farmer suicides, increases in unemployment …).

Figure 3.1: Agricultural yields and productivity should not be mixed up

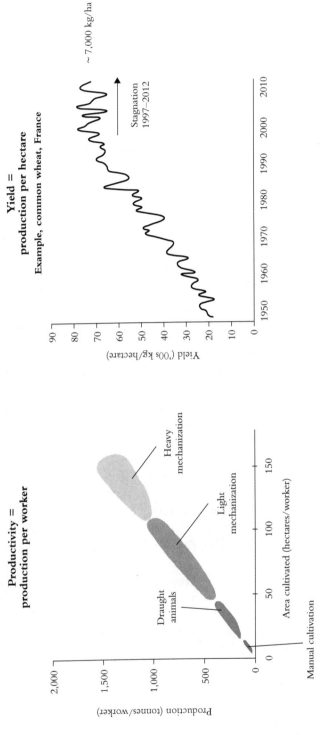

Sources: For productivity, Marcel Mazoyer and Laurence Roudart, *A History of World Agriculture: From the Neolithic Age to the Current Crisis*, New York University Press, 2006. For yield, Groupement national interprofessionnel des semences et plants (GNIS) [National Interprofessional Organization for Seeds and Plants]

The dubious promises of GMOs

Genetically modified plants (GMOs), their enthusiastic advocates often tell us,[1] are the way to feed a hungry planet while reducing environmental impact, and will be even more necessary to face the challenges ahead, namely to increase food production by a factor of 1.5 to 2 or more, according to the three parameters: population, diet and level of consumption. It is claimed they are able to do this by reducing the need to use pesticides (by producing their own disease or pest control molecules), increasing yields, allowing for the exploitation of poorer or more arid farmland that is currently neglected, and even improving the nutritional capacities of plants.

Is that really correct? Let us take a closer look at these promises. We'll ignore the case of vitamin A-rich golden rice because you need to eat several kilograms per day to obtain the recommended dose. What are the current GM crops in the world (see Figure 3.2)? They almost exclusively comprise different varieties of soybeans, maize, cotton and canola (rapeseed), which are either resistant to herbicide application (for example Monsanto's famous Roundup Ready) or have the Bt gene that allows them to synthesize their own insecticide, or they combine the two characteristics (known as stacked traits, where gene stacking refers to the process of combining two or more genes of interest into a single plant).

Figure 3.2: GMO crops in the world in 2018

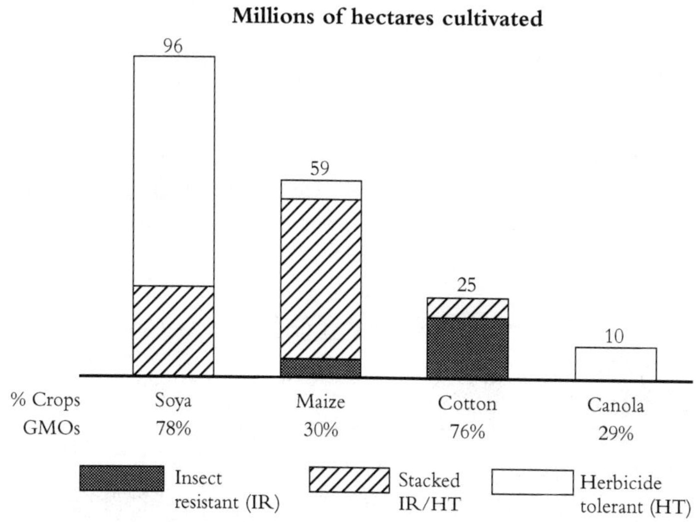

Millions of hectares cultivated

| % Crops GMOs | Soya 78% | Maize 30% | Cotton 76% | Canola 29% |

Insect resistant (IR) | Stacked IR/HT | Herbicide tolerant (HT)

Source: International Service for the Acquisition of Agri-biotech (ISAAA), Biotech Crop Annual Update, 2019

Without doubt, the Bt gene makes it possible to reduce the use of insecticides, but surely not to eliminate it. Cotton is a very delicate crop requiring at least 12 treatments per year to control lepidopteran larvae. Bt cotton can at best reduce the need for these treatments by 20–30 per cent. As for herbicide tolerant varieties, they are, by definition, made for the use of large-scale systemic herbicides, and in countries that have adopted GM crops there is an increase in the use of herbicides, which could be further extended with the emergence of weeds resistant to Roundup.

Finally, by how much do GMOs really increase yields per hectare? Plausibly, they can improve productivity (in human labour terms) by reducing the number of treatments to be made, and therefore the number of hours of work for a given area of crops. Yield might be better than for untreated land that would suffer crop losses, but surely not because 'it grows better'. As for GMO varieties that can grow in arid regions, none has been shown to be effective to date, while the necessary characteristics (resistance to drought or flood, adaptation to local climates, resistance to certain pests ...) often do exist already in very many traditional non-GMO varieties.

I am thus far from being convinced that GMOs are a response to the problems of a hungry planet, but rather a technique – and certainly quite a risky one – to produce more or less the same amount with fewer people, substituting machines for human labour and through the use of chemicals that do not biodegrade easily.

What directions?

If we take a step back from the current system, the solution to all our ills will appear remarkably simple. In the race for economies of scale (by applying industrial methods) and productivity (in human labour terms), we have been left with an almost complete split in activities in our agricultural practices.

On one side we have arable crops (cereals, fruit and so on) grown in monocultures in ever larger fields to allow ever-heavier machinery, increasing the risk of pests and therefore the need for preventive or corrective pesticide treatments. Soils that are over-treated and biologically 'dead' (with the disappearance of microfauna, fungi, bacteria and so on) are more sensitive to micronutrient depletion and water and wind erosion.

On the other side we have giant industrial farms, high-tech factories for chickens or pigs, robotic milking parlours for dairy cows, calves reared on slatted floors in crowded conditions, and so on, which challenges our relationship to other living beings,[2] and whose effluents are naturally impossible to manage since they are completely out of proportion to the land area over which they are spread – and let's not advocate making biogas, which can only be a second-best. Such farms increase risks from all perspectives: massive use of antibiotics to fight against inevitable epidemics, genetic fragility from over-breeding (entire beef or dairy herds from only a few breeding males, queens selected from beehives …), health risks and difficulties of traceability, and dependence on and economic exposure to soybean imports to supply proteins as a complement to maize …

The solution is therefore to accept a decrease of agricultural productivity by reducing the size of fields, and once again mixing and rotating crop, arable and livestock activities, as was always done until recent decades. In short, and to simplify, 'it was better before', when each household had a few laying hens, and milk had a different taste between winter and summer (as a consequence of the change of diet of the cattle between cowshed and pasture)!

Besides, when we talk about decreasing agricultural productivity, we must pay attention to what we are comparing, because in calculating agricultural productivity (so many 'farmers' producing so many tons), we should add in all the other jobs needed to run the system: agribusiness, industries, transport, infrastructure. The real productivity is not necessarily so great because the split between direct and indirect jobs has changed. Today there is around one agri-food job for each agricultural employee.

Using smaller fields would allow animal effluents to be returned to the land as fertilizer, which would be a partial substitution for synthetic fertilizers. The development of hedges and groves of trees would shelter small predators, especially birds, and, combined with the annual rotation of crops and polyculture, would reduce the risk of losses from disease and pests, and therefore the need for pesticide treatments to achieve a given yield. Soil fertility could be reconstituted in a natural way, alternating, for example, between cereals and legumes. Small farms would help maintain the genetic diversity of varieties and breeds and create more value added, more jobs per hectare cultivated, to combat unemployment and rural desertification (see Part IV).

As an added benefit, our agricultural landscapes would be all the more beautiful and pleasant to live in. If we compare the flat, featureless expanse of the Beauce, south of Paris, with the hedgerows and pasture of the farmland of Normandy, England and Ireland, we quickly realize the impact of our production methods. The number of farms in Italy is still 1.7 million, compared with fewer than 600,000 in France, although the latter has more than double the agricultural area. Perhaps this is part of the secret of the beautiful Tuscan landscapes?

Rest assured, I am not proposing that we return to the sickle and the scythe, with the mobilization of all available hands every year at harvest time, although it seems to me that the roots of school summer vacations are there. It is clear that, at least for arable crops, a good level of mechanization must be maintained, given the arduous nature of work in the fields. Nevertheless, we could intelligently reduce the size of fields without preventing mechanical work, by using long narrow fields to reduce the number of turns. The amount of fuel consumed by agricultural machinery remains modest in global terms, and by privileging this use, we will have enough to keep us going long after peak oil.

In some cases, 'de-mechanizing' can even increase yield. For example, in intensive organic gardening, harvesting by hand means the space between rows of carrots can be reduced, or several different vegetables can be grown together. Or fruit trees might be combined with the cultivation of vegetables or small livestock (keeping hens under apple trees, to eat caterpillars and at the same time reduce the need for chemical treatments), and so on.

Utopian, impractical? If there is one area that we could afford to change, that is it! The amount of agricultural subsidies that are distributed is such that all we need to do is to direct them towards compensating for the cost of human labour, subdividing rather than amalgamating fields and farms, and protecting farms near cities. Ideally, we would also take the opportunity to facilitate a transition to organic farming, but even conventional farming in such a system would need much lower inputs of treatments and would result in products that are less loaded with carcinogenic chemicals.

Nine billion and me

I hear critics say: "This is crazy, you want to starve the planet. High agricultural yields help feed the world. We cannot adopt organic farming as a widespread practice without a drastic reduction in yields." Well, let's see.[3] While it is true that such an approach is more intensive in labour terms, the yield per hectare is not inevitably lower and can sometimes even be higher, with appropriate choice of crop variety and good cultivation practices.

Nevertheless, it is clear that this would lead to a partial reduction of imports (the Brazilian or Argentine transgenic soybeans used to feed our livestock) and exports (there will be more local consumption in the arable/livestock balance) for some countries such as the United Kingdom, France or the Netherlands. Even if we reduce average yields and therefore the overall production, we have capacity in hand right here.

The first reason for this is the dreadful food waste in Western countries, since we throw at least a quarter (some sources even say a third) of our production directly into the bin. These losses are associated with the production and marketing constraints of the agri-food and supermarket production chain (such as promotional offers with specified expiry dates), communal catering with menu choices, and certain cultural aspects (for example the decline in cooking recipes that make use of leftovers).

A second reason is the balance in our diets between meat and vegetable calories: since it takes between 4 and about 12 or 15 vegetable calories to produce one meat calorie (4 for chicken and 12 for beef, which explains why more chicken than beef is eaten in poor countries, at least when meat is eaten ...). If necessary, it would be quite easy to reduce the overall calorie consumption simply by humans eating the corn and soya that today we feed to hens and cows. Clearly, this is not the current trend, with the emergence of a global middle class, especially in China and India, which is increasing the share of meat products in diets.

It will take a lot before we collectively go hungry in our Western countries, and we could accept some decline in yields, starting by properly finishing what is on our plates, learning not to spoil food, and eating (good) meat or fish more judiciously, perhaps two or three times a week.

The arrival of the next billion human beings should also not panic us, although it is of course desirable that the numbers climb as little as possible, for other reasons (energy, urbanization, transport ...). With the meat diet of a North American or an Argentine, no more than 2 billion could live on this planet, but with that of a Bangladeshi, we could happily be 12 billion. However, I am not wildly excited about the Bangladeshi diet, which is difficult to marry to the stressful, 'Western-style' life fond of fast calories and sugars:

> Man cannot live and work in a technological society unless he receives a certain number of complementary satisfactions which allow him overcome the drawbacks [...] the diversity of food, the increase in consumption of nitrogenous foods and glucoses are not a gluttonous overload but a compensatory response to the nervous expenditure caused by this technicized [sic] life.[4]

Those who allow themselves a small square of chocolate (or more ...) after a hard day's work can only confirm the accuracy of this statement. Could the more brutal capitalism of the nineteenth century have been possible without West Indian exports of sugar and advances in sugar beet chemistry?

The delicate question of returning nutrients to the soil

With livestock and agriculture once again mixed on the same land, we can usefully address the question of fertilizers, which can once again become largely organic: manure and slaughterhouse waste (bones, horn, dried blood) for nitrogen (N) and phosphorus (P), and ash for potassium (K) …

However, since Lavoisier's principle is still valid, there are nutrients present in the food we take from the soil that need to be returned, and which is not currently done, hence the compensation by synthetic or mineral means. So, I come to the delicate part of the 'programme', because the best way to recover the precious nitrogen, phosphorus and potassium is from urine and faeces. Regarding the latter, for a long time it was a habit of farmers and market gardeners near towns and cities to harvest enthusiastically the collected and dried 'powder' from urban excreta.

If we say that, at our latitudes, almost all wastewater goes through either a treatment plant, or a septic tank, there are two possibilities for collection of faeces, either 'at the source' (using dry toilets in individual housing and separate collection of urine in collective housing – there are examples in Northern Europe …), or by recovery of sludge at the outlet of the treatment plant. But, in the current system, these sludges are highly polluted: by the chemicals present in wastewater – cosmetics, cleaning products, paints, drugs – and by pollutants washed from the ground by rainwater, mainly from the exhaust pipes and tyres of vehicles. By reducing this pollution at source (using 'bio' cleaning products and cosmetics, and through a very large reduction in the use of vehicles in the city), we could reduce the contamination of this sludge, but undoubtedly not completely, because of the pollution already present on all artificial surfaces, which will take a long time to clear, if at all, when the bitumen itself contains traces of heavy metals.

The necessary approach will be different whether one is in a dense urban area, the suburbs or the countryside. But I already sense you holding your nose and I leave the question there. One last word though to those who might be disgusted by dry toilets: in our system of water management we have not evolved that much from the Middle Ages. Drinking water is taken from rivers (at least partly, and the rest from groundwater), and sewage is discharged. The difference is that there is now a post-capture treatment and one before rejection. But the downstream of some is the upstream of others, and unless you live in the mountains you literally drink the wastewater of neighbours who live upstream. We avoid intestinal disturbances thanks to chlorine, essentially. Is that so much better off than dry sawdust toilets? And don't just rush off and drink bottled water!

And for fish on Friday?

Fish are in the process of exhaustion and stocks everywhere are victims of overfishing[5]. As is the case for the yields of energy, there is one indicator that does not deceive: and that is the fall in CPUE – the catch per unit effort – which shows in the need to use ever larger and more powerful boats, equipped with high technology (for example, tuna purse seiner equipment for detecting and identifying schools of fish), and fishing in ever deeper waters (130 year old deep-sea fish end up on our plates), without increasing the global catch. In fact, it is the reverse, it is stagnant or slightly down at about 90 million tonnes per year.

Cheerful optimists tell us that, fortunately, aquaculture is growing! It will be the answer to the shortage of fish and the collapse of fisheries around the world. But there is a catch: we prefer to eat carnivorous fish, even the super-predators of the oceans – bass, tuna, swordfish, shark fins at 'trophic level' 4 or more. It takes 3–4 kg of wild fish to make 1 kg of farmed fish such as salmon or sea bream! And 20 per cent of the world catch, such as Peruvian anchovies, beautifully called 'fodder fish', is already devoted to aquaculture.... So, we should not expect a miracle there (except, by the way, through farming of vegetarian fish, as in the case of carp in China) as it will necessarily be a limited activity.

Regarding distribution and consumption

The disastrous aspects of large-scale retailing are well known: the proliferation of packaging, the excessive occupation of land (for buildings and parking lots) and use of energy (for lighting, heating, air conditioning and refrigeration), the dependence on the car, the pressure on suppliers, the distances between producers and consumers and consequent need for transportation – hence the need to return to shorter distribution channels, to more local businesses and markets. Markets have the great advantage in particular that the space they use is only occupied for part of the time, and for the rest of the time can be used for leisure and other activities.

An agricultural system based on smaller farms and plots, with production based on mixed farming, would clearly allow this type of development: mixed farming inherently reduces the need for transport (it is easier to produce a little of everything), although it will still be necessary from one agricultural region to another, with some seasonal variations.

To do this, we will have to discipline ourselves, probably through a mix of cultural, regulatory and incentive approaches. We will need to limit the consumption of out-of-season products. We will have to agree to pay producers a little more for our food, by modifying the split between producers and distributors (so the producer gets a fair share), but also by spending more, agreeing to pay more for quality meat but eating it less often. The evolution of the 'consumer's shopping basket' over the past 50 years shows that we have never spent so little on feeding ourselves, but with a deterioration in the quality of products. We will have to lower our expectation of menu diversity in canteens, refectories and restaurants. To allow them to serve quality dishes, prepared and cooked on site and not simply reheated, it will be necessary to reduce the number of dishes offered. The most ecological dish is the dish of the day ... as in the old days when we stopped at a restaurant and were offered the soup of the day, and not a menu with 15 different dishes, often frozen or industrially made and with a high level of additives.

Our food choices mean that we have now, and we will continue to have in the future, the agricultural landscapes and the farmers that we deserve.

Transport and cars

The challenges

> Left to itself, the car ends up destroying itself. The time that its speed gives us, it takes away immediately as it carries us elsewhere [...]. It takes us to the countryside, but we will soon fail to find in a 100km car journey the bathing pools or green landscapes that are waiting for us a five-minute walk away.[6]

A prophetic passage from the 1960s. Rural ponds and rivers where people can bathe near cities have disappeared, replaced by viaducts, highways and the no-man's land between strips of bitumen. It is now clear that the freedom created by motorized individual mobility has a high price, from the environmental point of view (emissions of greenhouse gases and pollutants, consumption of resources, paving over of land ...) as well as the societal (noise pollution, fragmented living, health impacts ...). There is no source of energy, now or in the future, that will allow all of humanity to achieve the average mobility of a North American, or even of a European (and certainly not using electric cars). The consumption of energy and metal is such that our only choice is to abandon the civilization of the car, at least in the sense that we use the word 'car', that is to say, an object of the order of 1 tonne in weight carrying an 80 kg payload in most cases. We are far from taking that path, with a global fleet of vehicles that passed the billion mark in 2010. The need for additional roads and car parks leads to the systematic disappearance of millions of hectares of precious agricultural land (especially in China nowadays), often the richest because it is located in areas undergoing urbanization in the plains and on the coasts.

How far should we go in questioning mobility? How could we maintain sufficient mobility considering the current constraints of town planning? Can we rely on a massive development of public transport?

The bicycle, the true 'clean car'

What can be done without a car, when there is no alternative public transport, and while we still have significant need for mobility (because our neighbourhoods have not had time to adapt), and journeys, such as from home to work, that we have to make? It is likely that more and more of our fellow citizens will be forced to ask this question, as they face increases in fuel prices arising from the sharing of a constrained resource between more motorists. In the long run, fuel poverty will not allow them to continue as they do today.

The first possibility is to give up or reduce that which is not essential: travel less far for holidays, or at weekends, be more selective about outings, and so on. It will be very hard and frustrating, and will not achieve so much: 15 per cent of trips are made for leisure purposes, and 85 per cent for 'economic' reasons, that is, commuting, business trips, shopping, visits to the doctor …

A second possibility, which does address the 85 per cent of journeys, is carpooling. It is a rather simple measure: of course, you have to start to talk to others in your locality, to organize a little, to be prepared to wait, to lose a little time and to accept additional constraints, but, in the face of rising prices and impoverishment, or based on ecological convictions, there are people already doing this. There are internet tools and apps available to facilitate contacts. Although undeniably very effective, this solution will, however, end up topping out at perhaps a 20–30 per cent reduction in total fuel consumption (which is nevertheless huge) so it can only be temporary.

There is a third and more radical possibility: riding a bike. These are by far the most energy-efficient vehicles, since in addition to the payload only a few kilograms are moved. They have the added advantage of being extremely durable and repairable. Of course, their use will be better for the young urban professionals who live only a few kilometres from their work, in city centres or near upscale suburbs, than for the working classes who are pushed ever further into suburbs and who have to travel dozens of kilometres, every day, even when it is rainy or windy.

The arguments are valid, and can be answered in four ways. First, it is only a suggestion for the direction to take, knowing that trying to maintain the status quo of a car-based civilization is doomed to resounding failure in the medium or long term. Second, it's not just for long commutes between home and work. Much car use comprises trips of less than 3 km: to buy bread, pick up the children from school, to take the elder one to her music class and the little one to his dance class. Journeys of less than 3 km are becoming easier, providing that roads are safe – which is the case when everyone is riding a bike and not driving – and we have the appropriate equipment: bicycles with baskets for shopping, a box for heavier loads and seats for young children. Europe is more fortunate in that, for reasons of historical development, its neighbourhoods are more amenable to cycling, at least compared with the United States, where people may have a hard time without their pickups, and are therefore prepared to desecrate landscapes to pump the last barrels of shale oil. But then, everyone has a cross to bear.

Third, enormous technical progress has been made – but at a low technological level – with developments such as electric bicycles, foldable bicycles that can be carried on public transport, and recumbent bicycles that offer the possibility of travelling long distances with lower energy expenditure while avoiding back pain. Fourth, let's go back to Ivan Illich's notion of an overall average speed.[7] If we just consider journey time, we go a little faster by car than by bike (say 30–50 km/h compared with 10 km/h). But if we add in the time that we needed to work to be able to afford the means of transport (initial purchase price then fuels, insurance, maintenance …), we can calculate a kind of overall average speed, and the bike then goes back to the top, because its cost is so modest.

Of course, this is all rather hypothetical, because it is difficult to see someone who spends 1 hour a day in their car and 9 hours at work, transforming their day into 4 hours of cycling and 6 hours at work. And that is without trying to get the agreement of the employer, who might not be so keen on such a pattern 5 days per week, 47 weeks a year, especially if the job itself is physical. But, for a moment, let's imagine that it is possible to organize ourselves a little better, and to develop a better working time balance, in parallel with the decline in our need for consumption. Consider the share of gross domestic product (GDP) that is fully or partially devoted to the car and its associated technical system.

First there are the various manufacturers: the car manufacturers themselves, together with the manufacturers in their supply chains, the manufacturers of the equipment needed to build and maintain the factories (robots, machine tools), the producers of raw or processed materials: steel, aluminium, plastics (polypropylene, polyesters …), artificial rubber, paints, glasses …

Then there are all the associated technical and commercial networks: car dealers, garages and service stations, breakers yards and scrap-metal dealers, landfill sites; the entire fuel supply system, from oil wells to petrol stations (with thousands of employees just to check and maintain the operation of petrol pumps), through exploration platforms, tankers, port infrastructure, refineries and pipelines. There are also public works and the associated equipment, underground car parks (with steel, cement, sand and aggregates …), road construction and maintenance companies, manufacturers of bitumen, road stone, traffic lights and other road furniture, painters of street marking; and all advertising expenses, with the associated consumption of paper and ink in newspapers and posters …

Finally, there are all the ancillary activities involved, such as inspection, regulation or dealing with traffic and incidents: radar, police officers and their equipment, manufacturers of breathalysers, hospitals, surgeons, nurses, physiotherapists, judges, insurers (all with their buildings and equipment); administration (vehicle and driver registration documents, traffic offences); the regular cleaning of buildings made dirty by pollution, the treatment of contaminated water (drainage systems), and ultimately, the cost of climate change; and even part of the duties of the armed forces, with their technical equipment, to secure our oil or material supplies …

So, what does this mean about the number of jobs that depend exclusively on the car? We might find the answer in one of two ways. We could start from the average expenditure per household, but this approach would lead us to underestimate a large part of the system (for example, everything that goes through taxes and feeds the budgets of local authorities, and government ministries of transport, health, justice, defence, industry …) and it does not take into account the labour intensity of different economic sectors (€1 of added value in an oil refinery and €1 in an insurance company do not necessarily create the same amount of employment). To determine an approximate percentage of jobs allocated to the car, sector by sector, requires a long-term study.

I would say that, at the very least, all this represents 30–40 per cent of the entire system. This means (given the number of unemployed) that, if we were able to divide up work between us appropriately, then once we are free of the car there would only be two or three working days left per week, just by saving the work generated by our current motorized freedom (see Part IV).

Personally, I would buy this right away, and this would be my personal weekly diary: on Monday morning I would ride my bike to work, irrespective of distance, season or weather. I would spend Monday night on a camp-bed in a corner of the office and then go home on Tuesday evening to start my 5-day weekend! All this without affecting, at least in the short term, urban planning and the current zoning between commercial activities and housing. Urban planning would then have to evolve gradually to adapt to new transport schemes.

Of course, my calculations are a little misleading, because the road system would still be needed for uses other than the car: bicycles and buses for personal transportation, and trucks to transport food and goods whose production has not been re-localized. So, it is true that there will still be need for a little more bitumen. But it is also sheer numbers that create the need for ever more complicated infrastructures: additional carriageways, separated by crash-barriers; interchanges, traffic lights and roundabouts, instead of simple 'give-way' junctions, and so on. Many fewer in number, our trucks could use the existing roads, even if it means letting some of them get overgrown. Part of the urban and suburban public space would revert to pedestrians and greenery.

Of course, my five-day weekend is a little optimistic, because there is a good chance that my wife would need to work from Thursday to Friday or from Friday to Saturday! Because, obviously, if we are to apportion our work better, it will be to allow production apparatus to be shared, not for everyone to work at the same time, otherwise nothing much would happen for the rest of the week. But no matter! All this is therefore a little hypothetical, I agree, but the purpose of this 'demonstration' is not to present the ideal solution, but to show that it would be possible, without much development of our neighbourhoods, to divide our time differently between travel, 'external' (especially salaried) work and 'home' work, based on more free time, which we could use to cultivate our vegetable gardens, to prepare the toothpaste for next month, or to make jam with this year's fruit ...

So, let's assume that there are a few cars left, at least in the early days. Not to transport our politicians (they will travel in the same way as everybody else, that will motivate them), but for ambulances, for slightly longer journeys, for the elderly or people with reduced mobility. But we should radically change vehicle designs. Consider, for example, bubble cars or quadricycles that are even smaller and lighter than the old Citroen 2CV – very lightweight, with puncture-proof tyres, made of low-grade steel or glass-fibre, with limited engine power (40 to 60 km/h maximum speed), without gadgets, without luxury or sophisticated safety systems (with lower protection people are less inclined to go fast). With such a machine, we could already make great progress: fuel consumption of less than 1 litre per 100 km,[8] fewer and more recyclable resources, easier road-sharing with bicycles (lower speed differential), reduced noise ...

and of course lower mobility than today, but still appealing to those who are most reluctant to cycle, and still tremendously superior to anything that humanity has known except for the last two or three generations.

Public transport: consider reducing train speeds

You will not want to hear this, but the total energy consumed in transportation depends on roughly three parameters: the distance travelled, the mass transported and the speed. Therefore, we don't have a lot of ways to reduce our footprint. It comes down to reducing speed, reducing mass, and ultimately reducing distance travelled, that is, making it increasingly time-consuming and arduous to use individualized transportation, while adapting our economy appropriately.

What about public transport? Isn't it much more environmentally friendly? Of course, trains, trams, metros and buses are more efficient, but only if they are properly filled. And it is important to distinguish their very different characteristics.

Even if their image is less 'green', buses are probably the most ecological and sustainable system by far, even if it is difficult to convince oneself of this from behind their exhaust pipes. They have low infrastructure costs, the flexibility to adapt to changing conditions and roads, minimalist technological content and robust, simply maintained vehicles. Trams partly share these characteristics, because once the first infrastructure investment has been made, their low speed does not place too much stress on the equipment or track, even if the 'technological' content of the latest models (such as digital display systems) could be reduced without too much discomfort.

Metros, light rail and suburban trains, using modest speeds, use much more infrastructure and maintenance costs (labour, energy and materials) are much higher. They are more 'fragile', require sophisticated and computerized systems for safety management, and rely on extensive technical networks to support rapid intervention by maintenance technicians, management of spare parts stocks and so on. These characteristics reach their peak in the case of long-distance, high-speed trains, because the constraints increase exponentially, in order for the track to support speeds of 300 km/h or more, while the tolerance margins decrease proportionally.

Public transport can be a good response to the gradual but partial disappearance of cars. For trains, existing networks are fast approaching capacity, and it will be difficult to reopen abandoned local lines, as many have already been built over. It is in any case already difficult for passenger trains, freight trains and maintenance activities to coexist on the same network. Buses can contribute because they are a simple and inexpensive system, and for this reason almost the only one that works in many poor countries. As for high-speed trains, since the energy expended increases with the square of the speed, in order to simultaneously drastically reduce energy and maintenance expenditure, it would be enough simply to reduce the operating speed – currently aiming to compete with air travel, which should in any case disappear, at least over distances of less than 1,000 km. If we maintain the status quo it may well be difficult to maintain high operating speeds, and they may eventually be decreased gradually to save on the costs of maintaining an ageing network of an over-indebted rail system, while other aspects of society will surely also experience some deterioration once the energy and materials resource peak is passed.

Construction and urban planning

The issues at stake

The building, public works and infrastructure sector is an incredible consumer of materials and energy. When it comes to products with a long lifespan, say 30 years, the consumption of materials is a priori less unacceptable than that of mass-produced consumer products, automobiles and new technologies, which rapidly become obsolete. However, this is not always the case, because buildings are also subject to trends towards a general speeding up in society: ever more rapid renewal of shops and offices, change of logos, signs and paintwork as a result of mergers and acquisitions or rebranding, reconfiguration of buildings to support changes in end use, spending of municipal budgets on 'visible' urban renovations on sidewalks and pedestrian streets …

The question of urban planning and our built-up environments is a delicate one. We can seek inspiration from many prestigious utopians, in the tradition of Nicolas Ledoux and Charles Fourier, but fundamentally there are now too many of us to live in the ideal city or in the utopian *phalanstères*[9] and we will not be able to 'transport the city to the countryside'.

We can, however, identify four major issues for this sector. First, reducing the amount of energy consumed by existing buildings is a matter of necessity. Second, as an issue of absolute urgency, we should stop building on greenfield land. Daily, we are destroying the last natural areas, reducing the ability of future generations to feed themselves, preventing wildlife from coexisting with us and hindering adaptation to climate change. Losing 1 per cent of the countryside every decade, what will our lands look like in a century with such madness, knowing that it is mostly focused on lowland and coastal areas, the best agricultural land?

Third, we must reverse the ongoing trend towards urbanization and concentration: even if today we also have attempts to 'return to the countryside', it is more a question of peri-urbanization – urbanization of former rural areas at the urban fringe. We must 'de-urbanize', but without sprawl. Large urban centres are too resource-intensive, because they lead to more transportation and more infrastructure. The stone staircase becomes an escalator, the 'stop sign' and pedestrian crossing a traffic light and a footbridge, the shop a shopping centre, the workshop an office tower. This de-urbanization should not be done by urban sprawl – it is not a question of fleeing the city for housing at the urban fringes, whose harmfulness is well known – but through a rebirth and revitalization of villages and small towns on a scale that allows them to be anchored in the landscape, with reduced transport needs and a sufficiently rich social life. Sceptics should be reminded that in the 14th century, before the Great Plague, famine and the Hundred Years' War, the density of the French population reached half its current level, that is 20 million inhabitants in 320,000 km² (vs. 68 million today, but for 550,000 km²) Certainly, there was less built-up area per person, and some no doubt slept in the straw with the farm animals.

Finally, let us drastically reduce the amount of construction, first by optimizing the land area that we use (usage can certainly be intensified, as in the case of schools that remain empty at night and during weekends and vacations), second, by reducing the need for new buildings and focusing mainly on reuse and refurbishment of existing ones.

Reducing needs

Here too, we have a lot of easily available options. We could start by stopping all (major) infrastructure works: roads and highways, airports, tunnels, canals, viaducts, bridges, high-speed railways, new port infrastructure. Apparently, and this is good, it may be the way we are heading anyway, but for economic reasons.... We already have much of what we need, and since bicycles and bubble cars will eventually take up less road space, the investment is unnecessary.

It is then absolutely essential to 'make do with what we have' for non-residential buildings, to find all possible solutions to avoid concreting or asphalting another square metre of land. Let us give priority to refurbishment of existing buildings and brownfield sites for new developments (there are, for example, good ideas such as setting up offices above commercial areas rather than in the middle of beautiful fields). Let us give priority maybe to investing in 'ugly' places for economic activities, so that we prioritize and preserve the places where we spend time out of work – on the condition that the time we spend there is less than today. Let us reduce our standards, our norms, concerning purely utilitarian places: why not accept offices that are a little less aesthetically pleasing, a little worn, with walls that are bare or at least not recently repainted. Finally, let us think in a more multifunctional way. Too many buildings are being underused: churches could serve well, from time to time, as concert halls, and not always for sacred music. Can we not generalize this type of sharing, for example with meeting rooms that could be used by companies during the week and by associations and sports clubs during the weekend or evenings?

Then there is the subject of residential construction. We are currently continuing to build at a rapid pace – and many will say too slowly, despite all the negative effects – as a consequence of three factors.

First, the population is growing, typically in Europe by about 0.3–0.5 per cent per year. Which, by the way, is not very much when we remember what we are told about the sacrosanct figures for growth. For a growth rate of less than 0.5 per cent, GDP per capita is de facto decreasing.... Hence on the one hand the encouragements from the state to maintain the birth rate (family allowances and taxation policies) and, on the other hand, the subtle management of immigration, so that it is neither too large nor too small. In a country like France, birth rate contributes about two thirds of the annual population increase, immigration the other third. Because GDP growth is already not brilliant with some 300,000 new inhabitants per year, we can imagine what it would be without.

Second, residential construction is increasing faster than the population. The number of households (and therefore of dwellings) is increasing, the average number of people per household is decreasing, and the building space per person is increasing. This concerns, in part, a desire for an increase in 'comfort' by increasing one's living area, but also the impact of societal change: the ageing of the population, the fragmentation of society, divorce rates, broken families, people living alone whether by choice or otherwise. These new household practices generate increased pressure on consumption: increase in the housing space per household (one room per parent), double equipment, transportation between the different places of residence.... But there is not much can be done about this, because we are not going to decree that we must love each other for life, under the pretext that it is a more ecological behaviour!

Third, and finally, the rate of second homes and vacant dwellings is increasing (at time of writing around 10 per cent of dwellings in France or around 3 million dwellings, against 1.7 million in 1975). This accommodation is used for weekends, leisure or holidays and is occupied an average of 30 nights per year. Of course, not all of them would be usable for year-round housing (think of ski resorts), but in many regions, new housing must be built while there are unoccupied houses or apartments.

On the thorny demographic issue, let us be satisfied with a simple observation. Like everything else, the population cannot continue to grow forever. At some point, the population will reach a peak or plateau, at least on this planet! It remains to be seen when and under what conditions. Regardless of demographics, the building floor area per person is increasing. Perhaps it is time to accept that we need to tighten up a little (or will we be forced to do so, for economic reasons?), especially if we do not want to constrain the demographic parameters too strongly?

Shared housing, which makes it possible to share certain living spaces, may offer some possibilities but requires a robust and long-term commitment. And if not, we will need to relearn the virtue of bunk beds in children's rooms. It will be painful, of course, although once again, it is not necessary to go very far back in the past to find much lower levels of consumption, without the apparent happiness of previous generations having been much affected. We have gone from the dormitory to the student room when we are studying. Maybe the constraints on interior living space could be compensated by a better environment outside, which would result from the end of the toxic civilization of the automobile? Because our urban public spaces have been ravaged, made unbearable, by noise, exhaust fumes, dust and hazards.... 'Paris is no more; destroyed not by Hitler but by Renault'[10]! Without cars, public space could be reinvented, for meals, leisure, children's games ...

Finally, appropriate economic and fiscal mechanisms will be needed to ensure that the constraints on land do not lead to increased rents to landlords, to the detriment of tenants who are finding it increasingly difficult to find housing. Perhaps taxation on rental income, the policy for low-cost housing (now you can keep your large apartment even when the children have left), and taxation on inheritance might all be reviewed ...

Praise the slow life: low tech = slow tech?

From the large complexes of the 1960s to the newest office buildings and cheap low-rise constructions, modern buildings are not always particularly beautiful, and many age very badly. While buildings built before the 20th century tend to acquire a certain charm if they are properly maintained, our current buildings, like many objects, join the world of the disposable, with structures intended to be replaced every 30 or 40 years at most, or even much less in cases where there is significant land price speculation, as in Hong Kong.

Of course, they are under significant economic constraints (it is necessary to act quickly and cheaply), but architects and developers bear a heavy responsibility: for damaging specifications (whereas 'ecological' prescriptions on materials can be easily made, for example favouring local species in the choice of timber), for characterless designs that deteriorate too quickly, for using poor-quality materials, and in too many cases for a most dubious aestheticism.

Could we not take the time to build beautiful, thoughtful and sustainable buildings and, inspired by what needs to be done for meat consumption, build fewer but better-quality buildings, designed to last and to beautify cities?

Consumer products, sports and leisure, tourism

Everyday objects

We have already outlined some principles that should govern the design and manufacture of everyday objects, to make them durable and repairable, resource efficient and less polluting. Broadly speaking, it is better to use our grandparents' coffee grinders in combination with an Italian coffee machine than the latest electronic and plastic espresso machine, using expensive disposable capsules. It will probably be necessary to use regulation to deal with technical obsolescence and to require warranty and end-of-life management obligations for products such as large and small electrical appliances.

Here again, reducing needs at source can and must be a major lever. In our countries, we already have so much stuff, in fields as varied as tools, DIY equipment, toys and books. Why is it still necessary to manufacture and sell so much, when some products are in theory almost indestructible? Of course, it is possible to break a plate or a glass, to damage a hammer or shovel handle, to puncture a football, to break or lose a toy. A book gets dog-eared from being read. But manufacturing and sales volumes are much higher than is needed to replace breakage or loss. Every year, thousands of copies of *The Little Prince*, *Great Expectations* or *The Catcher in the Rye* are still sold, while the number already printed and available should allow anyone to find them or dive back into them as they please, without creating waiting lists in public libraries. There are about 100 million 'Swiss army' knives sold per year. More than 6 billion Playmobil figurines have been made since their creation in 1974, 3 or 4 per child in the world, and probably very few have landed in poor countries. Even if the plastic has yellowed a little bit, there must be enough in circulation for every child in rich countries to organize a naval battle with the entire crew of a pirate ship (for those who are not connoisseurs, this is a classic in the range). What has happened to them? Mine are wisely hoarded at my parents'.... Our propensity to accumulate means that all these objects – and hammers, screwdrivers and key rings – are to some extent hidden from use, locked in kitchen drawers, garages or attics, until the owner moves or dies, whereupon they are taken to the dump or flea market. Would the world collapse if we no longer made hammers, but only handles? I'd be curious to see.

Some once durable objects, such as watches, pens, handkerchiefs, have acquired the status of disposable products. More than 35 million watches are distributed each year in France, including 12.5 million that are sold, the rest as advertising media, corporate gifts, or offered with subscriptions! It would be easy to do without them, to rediscover a taste for beauty and sustainability. Do we really need orgies of paraphernalia made in China, shimmering gift wrap with balloons, party poppers and paper hats with awful disposable cups for all our celebrations (birthday parties, school parties, retirement parties, and of course Christmas)?

There are other simple ways we could adapt our daily lives. Since my discovery of the chemical and/or metallic content of inks, I have only written with simple pencils for a few official papers and I almost never print in colour. And I try – sometimes unsuccessfully – to direct my offspring towards coloured pencils or pastels rather than marker pens. We could also manage the wear and tear of certain objects differently. I get annoyed when I have to part with one of my favourite shirts, because in the past it was only necessary to change the worn collar and cuffs. Such practices start to emerge hesitantly, for example with toothbrush heads ...

Hygiene products, cosmetics

In the kitchen and bathroom, we have already seen that the local production of ingredients that are less sophisticated than we use today, and the implementation of short production/consumption loops, would avoid the use of many chemicals. Even before this happens – until our pharmacists become apothecaries again, and train and hire to take charge of local production – it would be easy to reduce our consumption and the negative consequences of these consumable products. As an illustration, let us take a few examples that are immediately applicable.

A simple soap can be used instead of shower gel: not all objectionable chemicals are avoided (soaps, even 'organic' or 'ecological' soaps, are far from being exempt), but at least we avoid the packaging. We can manufacture household products for some simple uses. Chimney ash should work well to clean floors and even linens (it contains potash, which shares with phosphates and soda the alkaline properties that give detergency). There are several websites on the internet offering recipes with many variations.

We can back off a little from make-up and the many different skin creams. Not much has changed since women used bismuth salt as white blusher or black antimony oxide for eyelid make-up, both based on heavy metals related to lead metallurgy. Take the time to look at the composition described on the labels; you will agree with me. And with the massive arrival of nanoparticles, I'd look twice if I were you! You can replace shaving foam in a spray can by simple soap (with a little oil it may even be less irritating), again for packaging reasons, and to avoid greenhouse gases from the foam propellant. And either go back to the cut-throat razor or make the blades last a little longer by shaving a little less (every second or third day, as has for a few years begun to be fashionable) or choosing to be bearded.

Sports and entertainment

Like any activity using material items (equipment, clothing, shoes, balls …), each sport brings its share of waste or potential irritations. At the risk of making enemies, some sports are certainly more polluting than others: for instance, yoga is better than parachuting or scuba diving!

For example, simple space consumption is a significant parameter. In ball sports (see Table 3.1), it may be necessary to discourage golf or bowling and encourage table tennis and basketball! This is what Kant would have advised us if he had applied his *categorical imperative* to sports activities ('Act only in accordance with that maxim through which you can at the same time will that it become a universal law.')[11] Indeed, we would be in a bad way if everyone wanted to play golf (its 'democratization' has had an undeniable impact on the consumption of space, water, pesticides …), and we would be better off playing table tennis, or volleyball in newly mown meadows! Unquestionably, martial arts, yoga or dance are even more economical. As for energy, compare the pursuit of sports indoors

Table 3.1: Applying Kant's ideas to ball sports

Sport	Court/pitch surface area (m²)	Number of players	m²/player (***)
Table-tennis	70	2 to 4	17 to 35
Basket-ball	400	10	40
Volley-ball (*)	480	12	40
Handball	900	14	64
Rugby Union (*)	7,000	30	233
Tennis	600	2 to 4	150 to 300
Football (*)	6,500	22	295
Golf (**)	100,000	72	1,388

Notes:

* Sports that are shown shaded don't have to have dedicated/man-made pitches; one can imagine playing in open fields, after the harvest on newly-mown fields …

** A rough estimate, based on a 10-hectare course of 18 holes, with 4 players maximum at each hole. If that is not practical, I am open to all corrections on the subject. In any event we'll be fine playing table tennis, especially doubles.

*** To be absolutely correct, we should also include the duration of parties!

or in heated swimming pools with the possibility of swimming – albeit seasonal – in rivers and bodies of water free of their agricultural or industrial pollution.

As for professional merchandized sport, it fulfils three functions: first, and above all, it serves as a medium for advertising; second, it is undeniably entertainment (don't forget Juvenal's bread and circuses if you are interested in social harmony); third, it is an example, to be emulated, of healthy competition, of learning to surpass oneself, and so on. We should naturally try to avoid the advertising function. There remain the functions of entertainment and competitive emulation, which are acceptable as long as we consider the scale. All these ever larger stadiums, with stronger floodlighting for better television pictures, all these flights for major events, from world cups to the Olympic Games, both by the professional players and by the spectators.... Such an energetic debauchery, linked to 'long-distance competition', would be inconceivable in a more sober world, because how can we restrict ourselves in one regard if waste continues in another? Meetings between opposing teams, leagues and competitions are therefore acceptable if they are more local, between teams from nearby cities, at a regional level (within cycling distance!), and exceptionally national or international. The 'show' aspect of the events would necessarily reduce in quality, but we would gain in the dimension of emulation and of participatory pleasure, with supporters and players who know each other.

The ease of transportation today, and networking via the internet, are seen as great opportunities. Both have enabled unprecedented concentration in all areas of entertainment, sports, music and others. Famous musicians organize concerts for 100,000 people, celebrated football players and film stars earn indecent sums. A tyranny of over-centralization transforms us into spectators instead of active people. We listen to music on our phones, but do we sing more often? It is not at all clear. According to optimistic statistics, about 10 per cent of the population regularly plays a musical instrument: not very encouraging at a time of mass culture. Couldn't we absorb a little less content and produce a little more? Go less often to the movies and do more theatre? Download fewer video clips and make more music, increase the number of concerts, especially acoustic music made using 100 per cent renewable energy?

It is likely that the tyranny of over-centralization and global networking creates as much unhappiness as happiness, as much frustration as dreams, as much disappointment as good surprises. Let us take two examples. In the commercial field, the internet, with its intermediation sites, gives potential buyers and sellers access to almost unlimited opportunities. But, on the other hand, generalized competition almost systematically prevents a 'good deal' from being achieved. Whereas in the past the buyer or seller could find an attractive transaction (finding an item at a bargain price, or on the contrary making a good sale), the almost 'infinite' number of players on the internet ensures that the item will find exactly its market price. This is what economic theory says. No more bad surprises or scams, but no more good surprises either.

In the area of relationships, a universal competition also exists. Sexual competition, thanks to the ease of transport and communication, has also increased. Whereas, 'in the past', it was only necessary to 'defend' your beautiful wife or husband from lustful neighbours (and with all the combinations you like), today a beautiful girl or boy may be spotted, often at a younger and younger age, by model agencies keen to remove her/him from the local market and to make a financial killing in the global market.

Tourism and travel

I will spare you a sermon about the impact of aircraft, or cruise ships whose propellers stir up sand and overwhelm fragile marine life. Tourism, as it is practised today, is often awful, whether it is 'VAT tourism' (four days of shopping in Dubai where luxury watches are so much cheaper you can save the price of your flight), urban tourism introduced by the arrival of low-cost airlines (a spa weekend in Budapest, nice), ghettos of holiday clubs or even 'fair trade' tourism, where people travel by plane but support conservation efforts to compensate, and where they visit tropical wildlife parks but spread mosquito repellent or sunscreen, which are very harmful to aquatic fauna.

The only solution is to invent, or return to, a 'low-impact tourism'. To avoid creating a concrete jungle with low occupancy and use of buildings and equipment, we must accept local living conditions and food, accommodation with local residents or, at the end of the day, camping. We are seeing the emergence of excellent practices such as 'couch surfing' and house exchanges, definitely the best way to avoid contributing to urban sprawl on holiday. But all this is much less pleasant, in particular less relaxing and more difficult to put up with, as we get older and it becomes uncomfortable to sleep on the floor. Well, rather than have swarms of planes and boats vomiting up hordes of retirees, travel would become once again largely the preserve of young people, who could spend more time on it, for example during school or between jobs, travelling less quickly.

Luxury products

Finally, a word about the ostentatious consumption of the very wealthy. From a resource-saving perspective it is very difficult to defend – just look at the fuel consumption of luxury yachts – at least 1,000 to 1,500 litres per hour for a 45 metre boat – not to mention their manufacture (it is not just the bathroom that is made of precious tropical wood), and all the port facilities necessary for their use. Perhaps overall expenditure is relatively limited at a societal level, because of small numbers. But I wouldn't bet on it, if we gather together yachts, helicopters, private planes, racing cars, luxury materials, hotel suites and remote-controlled jacuzzis.

What is certain is that this consumption is toxic in the example it sets, because the desire to imitate leads to a cascade of desires at other levels of the social strata.[12] It is therefore unacceptable: of course, because of the waste of resources for the pleasure of only a few while others have to tighten their belts (but I am afraid that this is nothing new), but also and above all because of the ripple effect it has on society as a whole. A low-technology society would enable and would at the same time *require* a reduction in inequalities, which can only be good for the planet. In addition to some outright bans, reducing excessive consumption could mean charging high or even prohibitive prices for disproportionate uses, while prices for basic needs (the first litres of water consumption, the first kWh of electricity use) could be reduced.

I can already hear the objections: the 'luxury' sector is one of our European strengths, creating jobs and a trade surplus! But let's be careful of what we are talking about, because there is luxury and luxury, and we can in this argument ignore products whose 'luxury' is essentially a matter of clever marketing (let us say handbags ...), which are therefore not very different from our everyday consumer products.

Products whose luxury arises from using a larger and more skilled workforce in their production than equivalent manufactured products (*haute-couture*, quality cabinetmaking, and so forth), have nothing that is a priori incompatible with the principles we have already described (significant emphasis on workforce and on the quality and robustness of the design), except perhaps the rapid obsolescence that is linked precisely to their ostentatious nature (the evening dress or the shoes that will only be worn once or twice). On the other hand, there are products that are unaffordable in a low-tech world, by their very design but above all their societal uselessness, from yachts to limousines, from private planes to rockets for (future) space tourists. Similarly, products based on rarity (precious metal jewellery, diamonds, furs, for example) are often based on exploitation that is harmful to the environment.

Some direct jobs would therefore be at risk, as for example in the specialist yacht-builders of Brittany or Bremen. And that is not to mention the indirect jobs created close to the harbours that these same yachts use, like the security guards for hip night-clubs or the vineyards of Champagne. But before creating *jobs*, this is creating (additional) *work* and extra consumption of resources. We will return to the question of employment (see Part IV). In the meantime, let the yachts moor elsewhere.

Learning to make the right choices

In the longer term, the fact that we will have to reduce consumption of energy and raw materials will surely force us to make certain trade-offs. We will have to think about how to fight the effects of time, how to preserve the best of our culture for future generations. Recently, our belief in the longevity of digital media has been challenged. We now know that these media do not last so long physically, and in any case no longer than the duration of the technical means necessary to read them, in terms of machines or software – consider the example of floppy disks or cassette tapes. Many of the billions of digital photos taken each year will last much less well than the sepia pictures of our great grandfathers in uniform.

To keep everything will be impossible, because of our energy constraints. From ancient times to the Middle Ages, until Johannes Gutenberg, scribes, who could not copy everything simply because of lack of time, papyrus or parchment, had to make choices between authors and works, perhaps partly by default, and with the hindsight of centuries perhaps not always the right ones. Should heated library shelves be kept to store the truck-loads of books that are published today? In the days of Aeschylus, Titus Livius or Plautus, were the equivalents of today's airport novels or essays by telegenic popular philosophers published, without any record being kept of them? If published, would they have passed the selection test of the monastic scribes? And who today can judge what deserves to stand the test of time?

New technology, informatics and communications systems*

The issue: how to avoid throwing the baby out with the bathwater?

It is difficult to challenge strongly the development of electronics, communication tools, information technology or the internet. These have enabled access to an incredible increase in computing power for everyday applications, from sizing bridge structures to weather forecasts; to the speeding up and automation of tasks, generating 'productivity gains' in industry or allowing more precision in equipment as for example with computer numerical control (CNC) machine tools; the ability to communicate instantly from a distance, to store and share information, with all the good sides (advancing science, health, creating a universal knowledge base, of which the online encyclopaedia *Wikipedia* is the most emblematic example ...) but also the bad ones (the vacuity of information provided in real time without analysis, of anecdotes from the other side of the world).

But the development has its dark sides. Poor baby, we need to recognize that the water in your bath is increasingly dirty and that we are a long way from the fantasies of the dematerialized economy or of green IT. There is an enormous consumption of resources and energy and a prodigious generation of waste. We are seeing increasing negative social impacts of 'ultraconnectivity' and 'technodependence': the rupturing of society, the cognitive impact on children, the violence of some computer games .... We are inundated by services with zero or negative utility: advertisements,

---

* Thanks to Xavier Verne for his research and our correspondence, which resulted in the following co-written text.

spam, multiplication of viruses, the overwhelming mediocrity of content on social networks or video sharing sites .... There may be health issues from ubiquitous Wi-Fi and wireless transmissions (the impact of permanent exposure to different electromagnetic waves has not been completely assessed). We see the strategy of the major internet groups of ever-increasing control and concentration, and of generalized surveillance.

And the developments to come do not bode well for the future: ever more rapid obsolescence of equipment, the emergence of Big Data and its corollary, the explosion of servers and data centres and the whole world – trees, sheep, humans, shopping bought in supermarkets – 'chipped' by RFIDs (Radio Frequency Identification Devices) in the Internet of Things. This may make it possible to speed our passage through checkouts, but it is terribly Orwellian in its possible consequences. And that is not to mention the dispersive use of metals to make each disposable RFID tag!

The material aspect of new technologies

Of course, despite use of the term 'virtual', digital and even cloud computing, is in no way virtual: telephones, computers, servers, and other devices all have metallic content, much of which is poorly recycled if at all. More and more antennae, transmission equipment, transoceanic cables and the like are being installed, even while the available capacity is already enormous, to cope with or anticipate traffic growth. And if you believe that glass optical fibre is an example of a tremendous saving of resources compared to copper cable, keep in mind that this 'glass' is not just made of sand, but also of boron and of rare metals such as germanium (30 to 50 per cent of world production) or terbium which increases the refractive index and confines light in the fibre, while gallium is used for high frequency electronics.

The electricity consumption of IT and telecommunications equipment is increasing exponentially – in terms of the number of appliances and their individual consumption, and by the multiplication of data centres, which emit so much heat that they need to be cooled by air conditioning, or by installing them near a source of cold water. The most recent estimates regarding energy consumption converge to a range of 2,000 to 2,500 TWh, about 10 per cent of world electricity consumption.

As in other areas, the problem is growing and accelerating. Western homes have never had so much equipment – 6 or 7 screens per home on average nowadays, with ever shorter lifespans. Devices are increasingly complex (more resources), integrated and miniaturized (less recyclable). As evidence, consider an extract from the technical data sheet of a recent smartphone: proximity sensors (to turn off the screen during a phone conversation), light sensors (to automatically adjust the screen brightness), infrared (for the remote control, and to browse web pages without touching the phone), thermometer, hygrometer, barometer, accelerometer and three-axis gyroscope, magnetometer (wow!), two microphones (for ambient noise cancellation) and two cameras, multiple core microprocessors with clock speeds over 2 GHz, video at 60 frames per second and slow-motion capability also … and this in a device whose volume has been reduced, which makes it even less maintainable and repairable!

The volume of data being stored and exchanged is exploding. Between 2007 and 2017, the volume of data exchanged to / from data centres increased by a factor of 22. According to a recent IDC study, the 'datasphere' would represent about 33 zettabytes (thousand billion gigabytes), of which around 15 are stored in the 'cloud'.[13] This already means several thousand gigabytes per web surfer. Millions of hard disks are installed every year to keep pace with the rapid growth – which will not reduce with the Internet of Things. Indeed, all the big players of the cloud – Amazon, Google, Microsoft, and others – are launching projects one after the other.

Videos exchanged and downloaded, spam, or data to be used for your profiling: when you access a page (free of charge), servers calculate in about 10 milliseconds the advertising that will be presented to you. You visit a site, but, in the background, dozens of others are bought into play without your knowledge (as the saying goes, if you do not pay for a service, it is because you are not the consumer, you are the product sold). The development of the 'cloud' is based on the sensible objective, at least in principle, of saving on servers by optimizing their use and therefore their number and consumption. However, the storage of data might be done on the other side of the planet with all the additional traffic that it might generate.

What are the limits?

Information technology is obviously the field where the level of our ambitions is the most fanciful, where the usual assumption of infinite resources on the planet is most obvious. An example that is a little technical, but quite revealing, illustrates this well. IP (Internet Protocol) addresses, those that identify your computer or mobile phone, were until recently encoded in IPv4 using 4 bytes, or 32 bits. In practice, there are 2^{32} (2 to the power 32), or 4.3 billion different addresses. As we were coming near to the point where these are all allocated, a decision was made to switch to a new IPv6 addressing system, which requires software updates. In brief, some kind of Y2K bug,[14] but for specialists, but I won't bother you with that.

This new IPv6 system is now encoded using 16 bytes, that is theoretically allowing 2^{128} different addresses. This is a very, very large number, certainly enough to go around and potentially enough to assign a code to each object on Earth. In fact, more than enough, since 667 million billion addresses could be assigned per mm^2 of surface area on Earth, including the ocean. Even if some bytes are reserved for technical reasons associated with the ease of sorting data in the network, there are enough addresses to label each manufactured object and each living being.

Well that seems to be enough … Yet, even in computer science, physical limits catch up with technology. The empirical 'Moore's law' (see Figure 3.3), which states that the capacity of microprocessors doubles every 18 months or so, is coming up against a very physical reality: below about 5–10 nanometres, leakage currents make it impossible for transistors to operate. Of course, there will always be an optimist to tell us about the 'quantum' computer, another technological gamble: we'll see in a few years then.

Similarly, processor clock speeds have been capped – at around 3 or 4 GHz, because beyond that, energy dissipation becomes far too high – so that all that remains to increase computing power is to parallelize the processors. As a sign of the times, we are now finally trying to improve software performance. Until now, despite the immense increase in computing power over the past two decades, a computer still takes roughly the same time to start up, of course with more features (antivirus, software updates, internet connection, device detection …), but the user has not benefited much from the race for power. The mutual obsolescence between hardware and software works quite well. But come on, we shouldn't see mischief everywhere …

Possible solutions, huge potential savings

Will a day come when personal computers are low tech, resource efficient and as easy to disassemble as a bike, with simple, repairable or recyclable parts? It won't be easy. But, in the meantime, the current mismanagement has been so extraordinary that we already have the possibility of drastic savings of several orders of magnitude.

Concerning network infrastructure and access, the number of redundant physical networks could be reduced (see Part II), favouring wired access rather than the energy-consuming wireless wherever cable networks are

Figure 3.3: The end of Moore's Law?

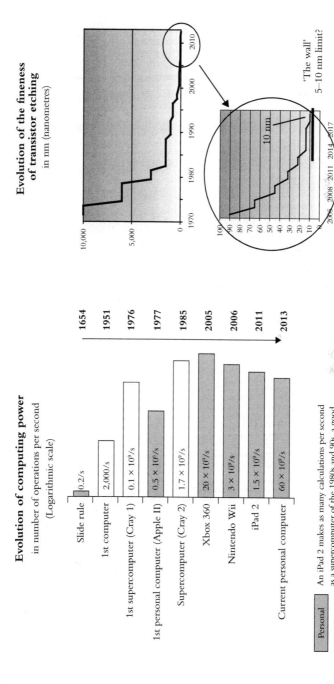

Source: Author's calculations, Wikipedia

already installed (ADSL, fibre optics). Mobile networks should be reserved for very limited use or in places without the infrastructure, such as in emerging economies.

Regarding servers and data centres, the waste, in terms of numbers and power consumed, is enormous. With appropriate re-design of the architecture, and by accepting the need for frugality, it should be possible to render the same services (file storage, web servers and so on) with one tenth of the existing infrastructure. The cloud may also have beneficial effects by pooling servers, provided that they do not generate additional long-distance traffic by preferentially hosting the data in the geographical area of use, unlike the current hyper-centralization. Today, internet giants are desperate for cooling for their data centres: Google installs servers in Finland (and Facebook in Sweden) to cool them with seawater, much like nuclear power plants. Green IT? OK, we save the energy needed for air conditioning, but is there not a better use for this electricity than heating the sea? Rather than setting up centres on a large scale, could we not decentralize them and use the heat dissipated to heat offices or housing, or to supply greenhouses?

With regard to end-user equipment, the potential savings are huge, first on the amount of equipment. Without going so far as to pool computers, for example by making them only available in media centres or libraries – it would be quite difficult to tear everyone away from their personal computers, although we could perhaps more easily start with printers – but do we need several televisions, computers at home and at work, tablets, mobile phones and landlines … not to mention cameras, camcorders … or even a Wi-Fi box per household? Without it being too much of a sacrifice for the geeks among us, we already have phones with the power of desktop computers, which could be used at home by connecting to a larger screen.

Then of course there is the question of the life of this equipment. Certainly, marketing plays a very strong role, to a point that always leaves me speechless, and technical obsolescence too. Sometimes to the point of absurdity, when the ever-increasing power and speed of PCs is ultimately translated into a loss of usability: software upgrades confuse users and may even prevent access to old documents. In business, a simple measure such as increasing the depreciation period of equipment would de facto curb the changes in computer equipment and avoid the scrapping of hundreds of thousands of computers. From this perspective, Linux systems, simple, resilient and less power hungry in operation, are increasingly used to extend the life of computers headed for the scrapheap.

It is possible that here the 'economics of functionality' may help to reduce the rate of renewal, provided however that operators are constrained by adequate regulation, otherwise the latest model will continue to be promoted by marketing. The first thing to do would be to prohibit the sale of subsidized devices with a monthly flat fee commitment. SIM-only mobile contracts have helped to develop the market for used devices.

Finally, there is a whole load of design as well as regulatory work that could be done on the devices to increase their lifespan and the possibilities of parts and materials recovery in recycling, but there will be things it would be necessary to give up: some aesthetic aspects, the race for small size and high performance (battery life, computing power ...). A modular approach could be adopted to promote 'repairability' and reuse of components: standardization of batteries, displays – which account for half of phone failures – sockets, connectors for accessories, maybe even processors ... and not just chargers, as the European Union (EU) has timidly imposed on some manufacturers! That would not prevent differences of design: in automobile design, product platforms often share up to 80 per cent of common parts.

There could be an enormous reduction in data traffic without much hardship, as so much internet traffic is of very little consequence (see Figure 3.4). Essentially, it comprises video downloading and streaming, spam and advertisements, attaching ever larger files without real benefit – consider the increase in photo resolution in recent years – consultation of sports sites, online gambling or continual *news*.... In the middle of all this are a few useful messages, some website consultations, some transactions on the e-commerce sites, some VoIP (Voice over Internet Protocol) conversations. Should we continue to install transoceanic cables to download even more Hollywood (or Bollywood) movies, Netflix series or appalling videos, at ever-higher resolution? The race for data capacity resembles the construction of motorways: the more capacity that is installed the more traffic there will be. By contrast, as we have started to do in limiting car access to cities, we should reduce capacity for users and services will adapt.

Of course, this would involve re-learning patience, giving up instant gratification, "I want everything right now (and free as well)." The entire *Wikipedia* text in English takes only 16 gigabytes,[15] the equivalent of four movies in DVD format! Even though the site is among the 10 most visited in the world, *Wikipedia* uses 'only' 350 servers in five locations spread around different countries, using about the same electrical energy as 200 US households.[16] The reason? Because the financial resources of the site are limited, it does not store high definition videos, it uses an optimized and intelligent architecture, and its mobile version is designed to optimize bandwidth.

Figure 3.4: Internet network usage

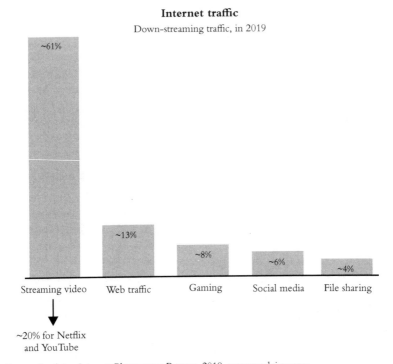

Internet traffic
Down-streaming traffic, in 2019

- ~61% Streaming video
- ~13% Web traffic
- ~8% Gaming
- ~6% Social media
- ~4% File sharing

~20% for Netflix
and YouTube

Source: Sandvine Internet Phenomena Report, 2019, www.sandvine.com

On the other hand, everything that is based on real-time interactions, from Facebook to Twitter, not to mention stock market operations that have out-and-out dedicated servers and transmission cables, is monstrously resource-consuming. Do we really need this 'ultraconnectivity', this cult of immediacy, to know at all times what is happening around the world, with continual updates and comments kept for 30 years? To allow this, the required infrastructure is huge: either the information is stored centrally, and we need to go to the other end of the planet to access it, generating enormous traffic, or it must be replicated on local servers, thus duplicating equipment.

Great progress can also be made with software, through eco-design, although this is still in its infancy. Server power consumption is not of sufficiently high priority for developers, whose tools mainly measure the response to user needs and pure performance, such as response time. But it is fairly straightforward to be able to improve the energy expenditure of a web page by a factor of 5 or 10 by following some simple rules.

Conclusion

Electronics are far more effective than anything else that has ever been invented to compute, communicate remotely, or share knowledge across the globe. Electricity and semiconductors are indispensable to the conservation of these communal goods. It is not a question of proposing a bamboo or organic computer, a rope Turing machine, or to give up our computing power and return to the abacus. But we should, as elsewhere, think about our real needs, whether with regard to videos or webcams, computation and high frequency trading (HFT) for the stock market, or weather forecasts! And, instead of fantasizing about an augmented humanity, we should think about the cognitive consequences of our use of digital devices. How can we not be sad seeing 30- and 40-something 'youngsters' in the subway playing Pokémon Go on their phones? Yes, life is hard, but even Huxley, in *Brave New World*,[17] had not dreamed of emptying our heads to that extent.

If we can significantly reduce the consumption of resources, perhaps we will not need to give up computers, at least not immediately. Well, we will see! But it will be necessary at some time or another to slow down the deadly and expensive innovation of this sector. After all, we have already frozen certain artefacts in their perfection: this is the case with the violin and with many other musical instruments.

Banks and finance

The banking sector is also eminently high tech. With the number of transactions it needs to manage and secure, it was one of the first to be computerized. It is enormously dependent today, and my intention is not to propose the replacement of cash machines by cashiers and the reappearance of gold or silver coins to replace the payment card. Even though that does not mean these will not come back one day, but that's another story.

The question of usury

We need to raise a more fundamental question than that of the consumption generated by the material activities of the banking sector, because it is the foundation of our economic system: the issuing of loans with interest, the 'usury' that was long prohibited by the Catholic Church (but largely circumvented by various devices, just as in Islamic finance today), but is now accepted as a virtuous practice and necessary to the workings of the system.

The fact of having to repay capital with interest implies the mechanistic and infinite expansion of the money supply. Since the total amount to be repaid, by all states, households and companies, is higher than what has been lent, they must all at the same time have access to a larger quantity of money. Moreover, with the creation of money by banks, central or private, through debt, it is necessary that the total debt also increases. Since the sovereign debt crisis, there is no problem in this respect for the moment!

However, if the money supply increases, the volume of goods and services produced must also increase, at least at constant velocity of money circulation, according to the Fischer equation $MV = PT$ (where M = money supply, V = velocity of circulation, P = price level and T = transactions). Otherwise, logically there will be an increase in prices, and so inflation, at the rate of money creation, let's say at the interest rate of loans. Seen from the lender's perspective, there is no more incentive to lend, since the interest recovered is cancelled by the effect of inflation.

It is therefore strictly impossible to avoid growth in an economic society that lends money at interest: stable production (let's not even talk about degrowth) would mean the collapse of the financial and economic system. So, should not we see the stagnation of production in Western societies as one of the reasons for the current financial crisis?

Eco-optimists will argue that we can continue to grow money supply and GDP, through green growth, by increasing production of services, but not of material goods. We return then to the issue of dematerialization of the economy which we have seen (see Part I) is largely fantasy. Services are, for the most part, tied too clearly to material production and consumption. To maintain an economic system based on interest-bearing loans without increasing material production, we need to develop services forever without increasing use of resources. This has begun to happen, of course, and many formerly free or even non-existent human activities are now sold, from tutoring classes to cleaning, from care for the elderly to dog walking. Much human activity is still outside of the mainstream economy, leaving some room for expansion. But who wants a world where we will be charged when our neighbours do us a favour or invite us to dinner? And there will be a limit anyway.

Money and debt in our societies

If we forget for a moment the absurd and rather esoteric language of economists and bankers, what is money in the end? It is the way to buy and sell, to divide two things: working time – either directly in the form of services, or incorporated into goods in the form of cost of labour (ah, the value of work …) – or the 'scarcity' of goods or services.

This 'scarcity' can be real or artificially created by our economic or cultural system, by our scale of values. It can be for renewable resources (animal and plant products …) or non-renewable (metals, precious stones …), movable or immovable property (land or housing). Lobster is more expensive than chicken, productive agricultural land more expensive than a stony field, gold more expensive than lead, for obvious reasons of availability, whatever the amount of work that we can put into play to get them. The intrinsic 'value' of goods or services can go beyond scarcity by mere availability: heritage value, historical content or rarity of the manufactured object. For example, an autographed photo of Elvis, a castle or an exceptional villa with sea views.

A debt is therefore essentially a call on the future work of the debtor (consumer credit), unless he has at his disposal an equivalent security to repay his debt (this is the principle of the mortgage, provided that the market does not play up too much). And fiduciary money, as its name suggests, does not have any value unless it is trusted, but also unless there are goods and services, and especially work, that it may be exchanged for at any time.

First aside: the futile debate about pensions

The debate over pay-as-you-go and funded pensions is a classic one. In many countries, public capitalization funds have been created or developed to help replenish and maintain the pay-as-you-go system. The capitalization funds are of course sensitive to some stock market crises. But beyond that, from a societal point of view the two systems are perfectly equivalent (but not necessarily from an individual point of view, although this is all subtlety). At a given moment, the same number of retirees expects (requires?) a certain number of goods and services from the same number of workers. Unless, of course, the funds put their money outside the country, in which case young foreign workers increase the number of working hours, but then – as is the case today, with the profits made abroad by multinational companies and pension funds – we are in a pyramid system that must also be based on global growth, otherwise it will collapse.

Second aside: why we will not dismantle nuclear power plants

Let's start from the assumption that the amount dedicated to be set aside today for the dismantling of our nuclear power plants is correct – although it is still widely debated, since the estimates of the necessary reserves varies typically from €200 million to more than €3 billion per reactor. This money has no material reality, it is about figures in a bank account, bytes, electrons on a hard drive. It is only a receivable, a claim on future human and material resources.

If it is not used immediately, and it will not be, since that is not the goal, it remains virtual. When the dismantling work is actually needed it will be necessary that society as a whole has the material means to carry it out: in resources (metals, cement, abundant energy), in technology (robotics, electronics, computers, transport systems and experts or specialized technicians), in motivated workers prepared to receive some level of radiation dose, if possible within the duly authorized limits. It will be necessary that the whole of the technical macrosystem will have been maintained, on the horizon of several decades or even centuries.

Nothing is less certain, as we have seen. So, I am betting – and I can do this because you won't come looking for me – that we will not dismantle anything at all. At most we will tinker a little in the early years, then, as and when the 'impoverishment' in resources of our society begins to bite, simpler measures will be taken, at first 'provisionally', and then the plants will finally simply be left in place, perhaps becoming future *taboo territories*, hoping that radiation barriers and maintenance of cooling water levels will prevent a dispersion of pollution.

The fundamental problem with nuclear power is that you cannot simply stop, and get out of it overnight. A refinery stops operation and closes, of course with some ripples and social dramas. A nuclear power plant, no. Cooling of the spent fuel needs pumps to be continually operated and monitored. In the United States, more than three quarters of all spent fuel is stored in swimming pools, which must be cooled permanently. In France, after the Superphénix reactor was shut down, it was necessary to keep 3,500 tons of flammable and radioactive sodium of its primary cooling circuit in a liquid state for years before gradually letting it solidify, still radioactive but easier to store.

Runaway effect

The banking system is not alone in creating a runaway economic system. This is true of all activities producing capital goods – machine tools, factory production lines, cranes, construction machines. Building cranes at a faster rate than they age or become technically obsolete increases the set of active equipment, and thus implies that there is downstream growth in the building sector.

The almost indecent health of Germany, at least at the time of writing, is partly based on this effect: thanks to the 'German model', to its highly praised and jealously guarded industrial small and medium-sized enterprises (SMEs), together (in particular) with Japan, it equips the rest of the world – with China in the lead and India close behind – with steel works, overhead cranes and automation systems, construction and bulk handling equipment.... But all this can only be done once, even in an exponential world. We will see. 'Brazil, Egypt are processing their cotton, Chile its copper. It is certainly still good business to sell to these countries weaving or wire-drawing machines rather than cotton threads or copper bar. But this market will reach its limits.'[18]

In a 'stationary' world it would only be necessary to replace machines when they wear out, without increasing their overall numbers. But for long-lived capital goods, production would be much smaller, and would probably call into question the economic viability of the manufacturing companies, or even the maintenance of the skills of enough people over time (we see today the difficulties of the nuclear industry in Western countries, for example).

So, what direction to take?

Let's go back to our problem of usury. This is a real issue, because even among the most virulent critics of capitalism, people have not yet thought to ban completely the interest-bearing loan. We can find keys to understanding in fascinating historical and anthropological books, such as Graeber's,[19] but they do not lead to concrete propositions for our present time. Without loans, how do we to ensure the functioning of the economic system? Even for low-technology activities, it is necessary to invest in production tools such as the baker's bread oven or the farmer's cart.

In order to come up with ways forward, it will probably be necessary to distinguish between three different kinds of loans: consumer credit, real estate credit and investment credit. Let's not spill any tears over the first two, which are determined by the way in which wealth is distributed today. That some people spend 30 years or more to become owners of their homes, while others inherit mansions, is pure social convention, and perhaps not very sustainable these days. But there remains investment credit, needed for economic activities. The answer could come from a complementarity between local financing using collective savings, as for example in tontines,[20] and public funding for investments on a larger scale, for equipment for networks of various sorts (such as utilities) for example.

To love, live and die in a low-technology age

To orient ourselves towards a resource-saving society, to rethink our technological world, would therefore lead us (will lead us?) to profound behavioural, cultural and moral change. Before concluding, I will present some thoughts on some even more tricky subjects. I hope that you will forgive me any errors as I depart from my field of expertise.

Love

Cecil Rhodes, founder of the De Beers company that today still dominates the global diamond trade, used to say that the future of his empire would be guaranteed "as long as men and women fall in love".[21] Can we change the values, the codes, the 'etiquette' of love? Give up gold and jewellery for the shell necklace, or at least make do with the huge existing stock? Only offer bouquets of flowers when they grow wild in spring and summer, and give up the dreadful cut flowers that arrive from Ecuador or Kenya by the planeload? Cut flowers are not just used, moreover, in the rites of love. I begin to dream of a world in which, when arriving at the home of friends, instead of bringing a bouquet virtually perfumed with kerosene, I could propose to the hostess to urinate in the vegetable garden to add some nutrients to the soil, thus increasing its future vegetable production!!

In addition to the direct consumption (of jewels or flowers), the search for a potential partner, together with the maintenance of social status, is largely responsible for ostentatious consumption, from make-up to sports cars, on which sexual competition is now partly based. Is it possible to profoundly modify our system of values and seductive behaviours, based on anthropological fundamentals?[22] After all, declaiming poetry does no harm to the planet, and, by singing under the balcony accompanied by his guitar, Romeo seduced Juliet with a completely carbon neutral activity ...

The delicate question of population

Population is the taboo issue par excellence. We have in the past been urged to "go forth and multiply" by religion but also, through their natalist policies, by states – previously to maintain numbers in the workforce and military, and today to contribute to the increase of the GDP. We will have to confront the issue one day, because of the question of the total population that the Earth can carry (depending on people's diets and the ability to transport food). Some parts of the world are already at their limits and, as Paul Valéry has written, 'We must remind the growing nations that there is no tree in nature which, even if placed in the best conditions of light and soil, can grow and expand indefinitely.'[23] What about the EU, with its 106 inhabitants per km^2? The average figure is obviously deceptive, since by removing the 'deserts' of the mountainous regions and the large agrarian areas, the concentration along the coast, around capital cities or in the low countries region is much greater.

In any case, we will need at some point to reconcile the irreconcilable, between the pleasure of making children and staying alive as long as possible, and our weariness with overloaded public transport or difficulty in finding accommodation. Purely as a mental exercise, let us imagine the moment when we will have reached a stable population, whenever that is, now or in a century or more, and whatever the level of population. At that point, we will have an equation of the type Births +/– Migratory balance – Deaths = 0.

Let's leave to one side the clearly thorny issue of migration (how can we deny people of goodwill the opportunity to earn an honest living, but how could we do it without "welcoming all the misery of the world"[24]?). The current compromise is based on a fragile and improbable alliance between militant defenders of undocumented migrants and employers keen for an exploitable workforce.[25] What is certain, net migration aside, is that to maintain a stable population, deaths need to compensate for births, or vice versa! So, the more progress we make extending life expectancy (which we are always very happy about), the more we will need to restrict the number of children we have. This perspective probably would have little appeal for our descendants as they would have to choose between a society where one lives to a (very) old age old and has (very) few children, or a society where one dies at a younger age on average but has a few more children.

Deep respect for the elderly is a common feature of all human societies. Old people – and we have always had them, even in prehistoric societies: life expectancy at birth and the average age of death should not be confused, once the critical period of high infant mortality is passed – had the longest experience, and were repositories of knowledge necessary for the survival of the clan or society. For example, it was better to know before building a hut to what level the waters rose at the last great flood.

Some resource-limited societies had their own original practices to 'manage the end of life', like the Kalahari bushmen:

> 'But these old people, how will they get on?' I asked [...] – 'They will go as far as they can', Ben answered. 'But a day will come when they can't go on. Then, weeping bitterly, all will gather round them. They will give them all the food and water they can spare. They will build them a thick shelter of thorn to protect against wild animals. Still weeping, the rest of the band, like the life that it asks of them, will move on. Sooner or later, probably before their water or food is finished, a leopard, but more commonly a hyena, will break through and eat them. It's always been like that, they tell me, for those who survive the hazards of the desert to truly grow old. But they'll do it without a whimper.'[26]

Or in Sumatra, from a text from the 19th century when the myth of the wildly cannibalistic savage was in full swing:

> 'Thus, today, the Battas no longer put their parents to death when age makes them useless as workers or as warriors. In the past, every year [...] the elderly were seen to submit themselves to suffering. The family assembled. Victims, worn down by age, collected all their energy, jumped up to a branch of a tree, and remained there suspended by their arms until, their strength waning, they fell to the ground. Then the children and the neighbours [...] rushed on them, knocked them out, cut up their bodies and devoured their muscles [...].'[27]

Of course, without going for solutions as radical as those of *Soylent Green*,[28] should we not think about the goals of the hyper-medicalization of the end of life, in a model where in any case medical and hospital expenditures are unsustainable? Why is it exceptional to die in bed? When asked by Croesus "if he had ever seen a man happier than he was", Solon, one of the seven sages of Greece, quoted two young brothers who "went to bed and did not get up the next day, but were found dead without having suffered pain or suffering, after having received so much glory and honour".[29]

Should you wither away in a hospital bed after three operations to keep you alive for a few more months? Why are we offended by assisted suicide for the elderly? Perhaps because we are a bit concerned about GDP. But perhaps also because our society feels some generational guilt? Because for the first time, in our world of the unbridled cult of youth – see, for example, those ridiculous telegenic professionals with endless careers, who want to look ever-young – old people are often no longer loved, respected for their knowledge and their wisdom (society has progressed too much and they are no longer 'in the know')[30] but 'millstones' that must be taken care of when they become bedridden. Or, conversely, because we are in a de facto gerontocracy, both political and economic?

Guide to an ecological death

Without wishing to be macabre, I take this opportunity to make some final (if I may say) recommendations on 'good practices' for funerals. Ecological funeral services are appearing everywhere, which perhaps shows that there is a niche market to be catered for.

First, regardless of religious considerations (although I suspect it will be more difficult in Varanasi), we must favour burial over cremation, for the simple reason of energy consumption, not to mention the emission of pollutants, such as mercury from dental amalgam. We take exactly the opposite path, unfortunately, because 'paganization' is gaining ground at least in some Western countries. We can then reduce the consumption of resources: we are starting to see coffins in certified timber (well, at least it's better than mahogany or another tropical wood), bamboo, cardboard, straw (that's better), without metal handles (you know my pet subject), with possible 'rental' of a wooden coffin so as not to appear too negligent in front of guests, who should, by the way, not be too many or come from afar. The ideal would be to be buried in a simple shroud of coarse hemp, at the corner of a wood, to boost the soil, but the civic authorities would probably object. In any case, why would we want to slow the decomposition rate of bodies with rot-proof wooden coffins and vaults in bronze – the alloy most resistant to corrosion? It also seems that bodies decompose less quickly than before, because of the food additives (preservatives) that we eat. I have this from a manufacturer of burial vaults, but I have never been able to confirm the information: I would be interested in any update on the subject!

Finally, it might be interesting to reflect on the space taken up by burial grounds. It's a rather pagan idea, but instead of a somewhat 'inert' place with neatly arranged gravestones and fine gravel crunching under the footsteps of the visitors to All Saints, we could have cemeteries that are 'sacred woods', of shamanic inspiration, smelling of good humus, and also fulfilling an ecological function as sanctuary for small animals and nesting birds. And the trees would grow very well thanks to the calcium phosphate of the bones.

When rubbish bins disappear

The most common word in France for a rubbish bin is '*poubelle*', named after Eugène Poubelle, the prefect of the Seine Department who in the 1880s made it obligatory for Parisians to be provided with a closed container for the storage and disposal of waste. This was unanimously acclaimed as a good example of the public health movement of the time. For some decades beforehand a debate had also raged between supporters and opponents of sewerage.[31]

Before these two great 'urban innovations', the streets of Paris were quite frankly filthy. The entrances of middle-class buildings had metal scrapers that were used to clean the soles of shoes, and it was quite possible to have the contents of chamber pots thrown over you while walking down the street. Paradoxically, however, the waste treatment system of the time provided a much more efficient mechanism than today for returning to the earth the nutrients contained in household waste and excrement. Excrement was taken from cesspits to the pestilential '*voiries*' of Paris – places that served as rubbish dumps from the Middle Ages to the 19th century – the most famous being that of Montfaucon (today in the 10th Arrondissement) where it replaced a notorious gallows. In the *voiries*, waste and excrement were transformed into '*poudrette*' (dried, deodorized night soil) that was very sought after by peasants as fertilizer. Household waste was simply thrown into the street, where it turned into a sludge (possibly after passing through the digestive system of a pig) which was taken by dustmen to the fields of market gardeners surrounding the city. The sewerage debate was not about the hygiene issue, but about the huge waste of manure that would result from the disposal of excrement into the river, at least before the sewers were directed to treatment plants outside the city.

Today our dear Monsieur Poubelle would certainly be shocked by the actual contents of his 'invention', and by our waste management system. Talking just about household waste, we produce more than a kilogram of garbage per person per day, all materials combined (industrial waste is in much larger volumes, as we saw in the Prologue), and only manage to recycle about 15 per cent (see Part I). Of the remainder, half is dumped in landfill, half is incinerated (or 'energy recovered' in Newspeak).

This mixes together valuable nutrients that should be composted and returned to depleted farmland, with materials that could be recycled (metals, glasses, some plastics) and other non-recyclable substances, often potentially polluting and difficult to eliminate (plastics, composites, with traces of metals in additives ...).

Millions of tons of all this is happily buried in landfill, or goes up in smoke, or is found insidiously in incinerator ash – the 'clinkers' that are themselves buried in landfills or spread as foundations for roads and parking lots, which allows them to be included in the statistics as being reused and value recovered. This is a scandalous practice given their content of metallic pollutants (lead, cadmium, nickel ...). But hey, as bitumen is to be spread on top, also containing heavy metals, why worry?

However, by examining the contents of our bins and given the reduction of waste that could be achieved by all the changes already proposed, it should be possible to do without bins once more! Let's explore this item-by-item (see Figure 3.5).

Putrescible (decomposable) waste makes up around one third of the contents of our bins. The solution obviously involves composting, but adapted to each situation: for separate houses through individual composting (also to be encouraged to reduce the weight of material collected, itself a good thing), in dense urban habitation through individual 'vermicomposting' – which I can guarantee is odourless, I tested it in an apartment for several years, but it is necessary to put your fingers in it from time to time – or composting in the courtyards of shared buildings. In this latter case concierges could be entrusted with the operation of the composting and could make their return with the help of garbage collectors turned into 'compost consultants'. The compost could then be resold locally as fertilizer or used on shared gardens or fields in the area.

Even better, before entrusting our vegetable peelings to the good care of bacteria and composters, we could usefully raise farm animals (pigs, chickens, ducks, rabbits) and contribute to the necessary deconcentration of animal-rearing activities. There is nothing to stop you engaging in these activities even in the most urbanized areas, provided regulations and our system of values (regarding odours maybe) are changed. Why not a pig in each building courtyard, instead of the place for garbage cans? The residents would come down with their peelings, concierges '2.0' would be responsible for watching over them – the animals, not the residents! – policing would be transformed (with the disappearance of cars there would be more time) into vets able to make preventive visits, then a truck (or a cart, for ultras) would collect to avoid having to kill them in the courtyard ...

Incidentally, the pig would be a useful replacement for traditional domestic animals, such as the urban dog (ravenous consumer of food twice a day, and for conscientious owners, of small plastic bags also), or the domestic cat (serial killer of dozens of birds, failing to distinguish between sparrows and endangered species, the rogue). Faced with resource constraints, it would be better to bet on more 'useful' animals, such as sheep to keep the grass short or bees that produce honey and offer a 'pollination service'.

Figure 3.5: Household waste ... and its possible future

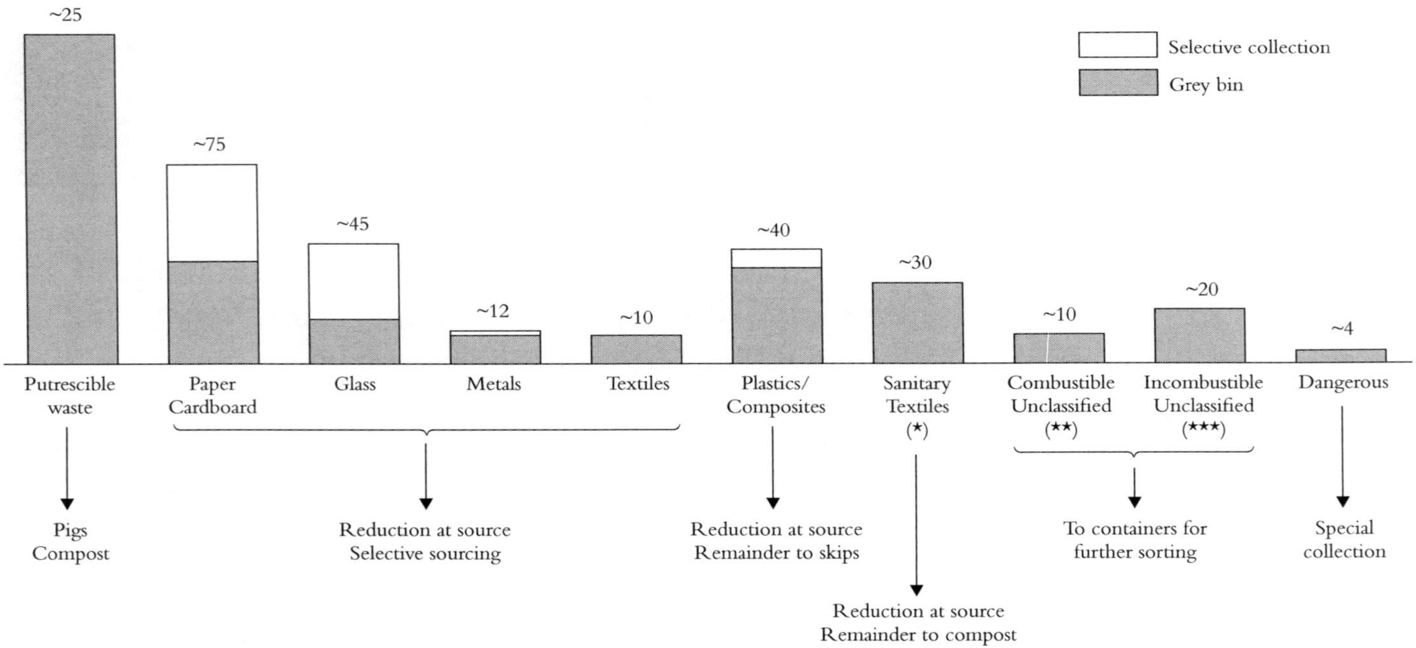

Legend:
- Selective collection
- Grey bin

Categories and values:
- Putrescible waste ~25 → Pigs / Compost
- Paper Cardboard ~75
- Glass ~45
- Metals ~12 → Reduction at source / Selective sourcing
- Textiles ~10
- Plastics/Composites ~40 → Reduction at source / Remainder to skips
- Sanitary Textiles (*) ~30 → Reduction at source / Remainder to compost
- Combustible Unclassified (**) ~10
- Incombustible Unclassified (***) ~20 → To containers for further sorting
- Dangerous ~4 → Special collection

Notes: * nappies, sanitary towels, wet-wipes, handkerchiefs, kitchen towels, paper tablecloths and towels ...; ** wood, leather, rubber, tyres, carpets ...; *** clay, stone, terracotta, gravel, shells, plaster ...

Source: ADEME, National campaign for the classification of household waste, 2007; ADEME Report, 2012

For paper and cardboard, glass, metals and textiles, there are already specific waste streams that should be used. There is still work to be done when we see the percentage of glass or paper that currently gets put into non-selective bins, but reducing consumption would in any case significantly reduce the amount circulated.

Plastics and composite materials are hard to treat (there are identification and sorting difficulties, with zero recycling possibilities for thermosets and limited possibilities for thermoplastics), as evidence of the difficulties in increasing the amount recycled, even though most products proudly display (to relieve our conscience) the recyclable logo: there is unfortunately a big gap between being recyclable and actually being recycled, or issued from recycling. The solution should come largely from a reduction in need (of packaging in particular) and from the re-design of products (to be simpler, with more use of single materials) as already described. A small amount might remain to be collected in neighbourhood bins to be either sorted or processed.

Then come the delicate sanitary textiles. The use of these has increased sharply in recent years with the introduction of household wipes (who do we have to thank for that?). Given the ease of doing without, and that they are characteristically societally useless (the principle of why wipe yourself becoming why wipe?, as noted in Part II), we can easily deal with these, and with paper towels, disposable tissues and kitchen towels. Then we only need to consider nappies, sanitary napkins, cotton wool.... There are very effective alternatives, such as washable ones, but for nappies this requires a significant commitment in our current society, but could be simpler to use if everyone adopted them (there would be a critical mass to allow for pick-up/laundry services for example). The other way would be to ensure that these products, if they remain disposable, are composed entirely of biodegradable products, which would mean giving up part of their absorbent capacity, because the excellent liquid absorption capacity is related to the polymerization of a chemical based on acrylic acid.

Finally, the rest is composed of many materials, partly inert (terracotta clay, when you throw away your broken flower pot ...), partly recoverable – recyclable or reusable – provided you can recover the material (rubber, leather ...), partly dangerous, polluting or destroying. One could imagine skips, perhaps of several sorts, which would be sorted later. Some skips, destined to receive valuable products, would be 'scavenger bins' (yes, rag pickers would reappear, another step backwards and a return to the candle!). Other materials may be put forward for destruction by incineration, but in much lower quantities than today, say a few kilograms per capita per year at most.

It would therefore be possible, and this is not utopian, to get rid of bins eventually. So, let us dream of the blessed day when the trash will disappear. On that day we will be able to return to the earth all the nutrients extracted by agricultural production. Because, on the excrement side – and I am forgetting my dry toilets for the moment – sewage sludge would, unlike today, no longer contain many pollutants (heavy metals among others) which make it potentially problematic to spread on agricultural land. On the one hand, the disappearance of most cars and their particulate emissions would mean that rainwater would no longer be loaded with heavy metals and, on the other, the reduction in use of certain chemicals and metals should significantly reduce the pollutant content leaving wastewater treatment plants.

Of course, there will still be specific waste treatment streams, such as hospital waste, which cannot be treated other than by incineration (perhaps some use of disposable items might be reduced, but by much more frequent use of washing machines at 90°C?). But, in any case, the quantity is tiny compared to household waste, 200,000 tonnes compared with 28 million in France. Finally, it should not be forgotten that household waste is only a small part of the waste generated by the activities of businesses (90 million tonnes in France, excluding agri-food and building and construction). But reduction in production and activity should automatically reduce these quantities as drastically as in the household sector.

And where is energy in all this?

Let's take stock of the necessary transition

How might the energy issue be solved in our low-tech world? We saw, when discussing the subject of network industries (see Part II), that the production and distribution of energy are also devilishly high tech.

On the distribution side, currently a set of computers and automated controls triggers additional production capacities (gas plants, hydroelectric dams) to adapt production to consumption peaks. That will work differently with the arrival of smart grids and individual 'smart meters' that may, one day in the future, track demand and switch off devices (temporarily disconnecting your washing machine or an industrial motor) to better adapt consumer demand to available production.

On the production side, let's not talk about nuclear at the unmanageable technological level. But then again, apart from the risk of accident and contamination (which makes all the difference, of course), some a priori 'green' renewable energies, such as large-scale wind and solar photovoltaic, are not very different in terms of technological content and technical complexity. There is not much difference between a nuclear power station and an industrial wind turbine of 5 or 7 MW, or rather, a macrosystem of thousands of wind turbines and photovoltaic 'farms', connected by smart grids allowing intermittent supply and variable demand to be balanced at any time.

Can we imagine a 'stable' world based on these technologies, maintaining such a system for centuries or millennia? Of course not, and for the same reason as for nuclear power plants: it will be necessary to renew these large wind turbines every 30 or 40 years, it is impossible to recycle all materials properly – that is to say without degradation of use, specific critical resources will be missing, and there is a hidden dependence on fossil fuels (for metals, plastics, polymers) and on the entire technological *megamachine*.[32]

So, what is the point of embarking on a major industrial renewable energy programme to maintain our current level of energy expenditure (or close to it, with energy efficiency), if this is not sustainable in the long term? Some people answer that it is to buy time, to help us make a 'transition' over one or two generations. But in that case why not organize the transition *right away*? This would help avoid wasting precious resources – energy and materials – in the interval as we continue to speed up the present system.

Good use of (really) renewable energy

We will then need to recognize that truly sustainable forms of energy are undoubtedly those based on systems that are less 'aggressive', more local, adapted to their environment and therefore, we come back to it, relatively low technology – in order to be achievable, repairable and replaceable locally – even if this means sacrificing a little performance and efficiency. This means micro and mini hydro-power (and not too much, because its impact on fragile ecosystems of rivers can be very damaging), small 'village' wind turbines, solar thermal for health and cooking needs, biomass and biogas, possibly supplemented by heat pumps …

Unfortunately, the amount of energy recoverable by such technologies will be low compared to our current Western standards. It will not be worthwhile to try to run numerous escalators, high-speed trains or large chemical (chlor-alkali) and electrometallurgical (aluminium, steel) industrial sites. But if we organize well, there will probably be enough to live decently and not return to doing our laundry by hand. Although we will have to climb stairs, reduce the speed of our trains and give up aluminium drink cans.

This is, by the way, another argument for 'de-urbanization' (see p 99), to return to villages and towns at the scale of a few thousand inhabitants, because while small wind turbines on a roof may be able to run some washing machines, they will not provide power to a building of ten floors or more! Not to mention the use of firewood. We should also think of slowing down – societally, culturally – to find the time needed for sharing of community facilities better and decreasing energy consumption. And, staying on the subject of laundry, it may be necessary to run our washing machines when the wind blows (massive use of batteries is a very bad idea), and therefore we should not be too far from home when that happens! After all, not too long ago, when we used windmills, I guess the miller had different activities on calm days ...

Ultimately, how much energy might we produce using these more basic technologies? It will depend a lot on the local geography. Also, are we talking about primary energy or final energy, the difference between the two depending on the technologies used (delivered as electricity or not for example) and their performance? I'll spare you the boring calculations.

If I had to try a number, I would bet on 20 to 25 per cent of our current (Western) consumption, at best, mainly in hydroelectric form by developing, in addition to existing dams, mini and micro stations (the return of water mills, operated either with electricity generation or direct use of mechanical energy to avoid generator and motor losses), with the addition of firewood, a strong development of individual solar thermal and small and medium power wind turbines, and occasionally facilities exploiting biogas.

Is it impossible to reach such a low level? That remains to be seen. The level of ambition is not really so shocking: as evidence, the very serious scenario proposed by the négaWatt Association (which does not really advocate a challenge to our 'comfort') that shows a halving of the final energy consumption in France by 2050, driven by sufficiency (60 per cent) and efficiency (40 per cent), despite a population growth of about 10 per cent.[33]

As we have seen, drastic reduction in energy consumption would not in fact be so difficult, working essentially on three types of levers. In the residential and tertiary sectors, on consumption for heating, this can be done by pulling on a sweater and/or by launching a major programme of thermal renovation of buildings, but with the limiting factor of the number of existing buildings (see Part I), because we cannot realistically renovate a million homes a year. In transport, by reducing mobility needs and changing modes of transport (across the whole range, from carpooling to cycling) and reducing the weight, speed and therefore the energy voracity of the remaining vehicles. In industry, simply by decreasing needs, through reduction in consumption of finished products (prolongation of the life, reuse) and by recycling.

Naturally, it is only a back-of-the-envelope calculation on the achievable level of ambition, and a precise projection, especially on the question of the oil/gas/electricity energy mix, which is vitally important, would be needed in each country or region.

PART IV

Is 'Transition' Possible?

It is time to reflect on the real possibilities of turning our crazy ship around, and of effecting a transition from our society in peril to a world of low technology that is parsimonious in its use of resources. While the proposals put forward might indeed be possible technically, we need to be very sceptical about the real possibility of making such a turn, or, to mix our metaphors, of applying the brakes to our fast-accelerating world.

This is clearly due to the severity of the necessary changes, to the systematic reversal of trends that have been in place for decades, even centuries, that seems so shockingly reactionary to the more progressive among us. But it is also because we have not addressed any of the cultural, societal, moral or political aspects that would necessarily accompany the technical and organizational developments: a fresh relationship with work, and new, more moderate consumer practices.

Is it possible to generate enthusiasm for such a 'programme', or could it at least become 'socially acceptable'? Could it be compatible with the various pressing democratic imperatives that require rapid results under penalty of immediate sanction in the polls, or with international relationships or the pressure of a media system fed by advertising revenues, and therefore inherently opposed to any form of sobriety? All of this in a world facing financial, geopolitical and climatic upheaval? The least we can say is that it will not be a done deal.

The impossible status quo

Let us assume that a transition of such complexity is indeed completely unimaginable, even if proposed and then supported by a sufficient majority. But, on the other hand, can we afford to keep going as we are?

Some possible analyses of the situation

Let us imagine the views of a random participant in a discussion on the current and future difficulties of the world: there is a good chance that a sensible person will recognize soon enough that things cannot go on like this. We all know this more or less implicitly, perhaps not admitting it so willingly because the cognitive dissonance with the need to continue to live on a daily basis, to make ends meet or to enrol the little one in the nursery, would be too great. It is even true for politicians, even for managers and engineers, who accepted and 'organized' the loss of many jobs and activities in their countries, before finding their own jobs at risk.

Opinion polls show it: the terror of the middle classes is 'decline', the mortgaged future of their children. After generations of technical and social 'progress' in Western countries, everyone realizes that things have changed and that life will be harder for future generations: to find work, housing, a decent 'place', hope ...

Each of us makes our own analysis of the situation. For some, full of confidence and enjoying privileged media access, the goal is to reform a failing Europe: it is not liberal enough, it's too timid in the face of scientific progress, too social. Historic privileges are locked in and growth needs to be revived, in order to compete in the face of fierce global competition, to become a country of winners that attracts capital and globalized talent. Of course, we all know that the grass is always greener elsewhere.... However, this view is rather short-sighted, drawing on a simplistic view of political economy and lazily extrapolating from past trends to predict future directions.

Others have taken a more questioning approach and have made a number of analyses of societal 'crises' that help to shed light on different aspects of a complex, difficult to grasp reality. They all seem to me to contain interesting and relevant elements.

The 'Marxist' perspective considers that the fruit has been rotten from the beginning. With rates of profit declining in the late 1960s, it was necessary both to implement a monetary policy that deliberately generated unemployment in order to maintain downward pressure on wages, and to encourage increased household debt to maintain sufficient demand, and therefore production. However, household debt reached an unsustainable level and the government had to take some of it on to keep the system afloat. See you at the next episode of the sovereign debt crisis.... This theory has the merit of matching macroeconomic figures rather well over the past 40 years.

Another, more 'degrowth'-oriented reading adds the saturation of our societies with manufactured objects (the household equipment rate) to the decline in purchasing power owing to debt or unemployment, which inevitably slows demand, despite the efforts of engineers and advertisers to encourage the technical or cultural obsolescence of products.[1]

Meanwhile, according to the 'peakists' who analyse oil and gas production peaks (past or to come), our major problem comes from supply constraints, the tension created by ever less accessible energy resources. Some economists believe that the subprime housing crisis in the summer of 2008 may have been triggered by the surge in oil prices at the time, undermining indebted households unable to both fuel their cars and repay their mortgages at the same time. Whatever the case, the decline in energy efficiency affects the 'wealth' of our societies, because the net production available for household consumption is declining, while an ever-increasing proportion is devoted to the production of intermediate goods. The hamster's wheel is spinning faster and faster, but without any real benefit for the end consumers.

In the same vein, followers of Joseph Tainter[2] believe that industrial societies eventually reach such a level of complexity, hyper-specialization, optimization and flow of materials and information that the system becomes unstable to the point that it can collapse relatively easily. Examples include the effects of Fukushima, which significantly disrupted the electronics industry, or the eruption of the Icelandic volcano (I lack the courage to look for its exact spelling). One can only be struck by recent economic difficulties of trying to support increasingly complex and expensive infrastructure, such as road and bridges, or hospitals – the issue is, are they economic or systemic? It is likely that, to function properly, our world requires a pyramidal structure with a large middle class of docile and imitative consumers.[3] It is impossible to fly a private jet in a world composed exclusively of proletarians. It takes a middle class, with its own dreams, to design, manufacture, maintain and refuel it: a whole world of engineers, mechanics, technicians, geologists and welders …

Finally, critics who take a development perspective[4] analyse the Western crisis at the end of the 1970s in the light of the growing imbalance in North–South relations, the deterioration of the terms of trade, the debt deadlock (even at that time) and the historical and persistent collusion between the state and capitalism, and between elites in the North and the South.

The impossible status quo

Whether or not we share these analyses, we do know about the state of the planet and the warning signs of potential or actual major problems.

Even climate sceptics cannot deny the warning signs on *all* the other parameters. Of course, radical change is very complicated and risky, but isn't the status quo worse? In addition to the physical danger of a possible 'collapse', we risk going individually and collectively mad in the face of incessant conflicting demands.

Consequently, our political and economic elites are moving, but in a rather uncoordinated way. In the morning, they try to revitalize industry, or at least to slow down offshoring, and then in the afternoon during a visit to a coastal city they call for the revival of ports. Understand that if you can, since the health of ports depends above all on container imports from Shanghai and elsewhere.... In any case, the real or perceived acceleration of change in the world puts a stop to any attempt at political and, increasingly, economic planning.[5] In many areas, we can see large companies preferring to buy competitors in order to acquire patents, production capacities or skills, rather than embark on a risky adventure due to lack of clarity of the situation.

As for us, we have become super-demanding consumers. We want quality, efficiency, responsiveness, availability, speed, helpfulness, everywhere and all the time, all at the lowest possible prices. But we forget that, at the societal level, what we demand as consumers impacts us as producers. Yes, industrial society has become terribly efficient and adaptable, but what we have left behind is visible in offshoring, unemployment and all the studies about workplace stress ...

Of course, not everything was better in the past. But what will the world look like in 30 years' time if we continue the changes of recent decades at the same pace? And how can any normal human being resist such pressure without despair?

Between wait-and-see, fatalism and 'survivalism'

In short, with a little common sense and curiosity, it is not difficult to come to the conclusion that 'we are going down the pan'. It takes a serious dose of optimism to have a reasonable faith in the survival of the Amazon forest. I'll bet that even the most dull-witted economists or progressives understand this. As evidence, even the nuclear industry, which (despite the failure, up to now, of fast breeder reactors) promises us thousands of years of abundant energy with breeder reactors or fusion (a long way ahead), at the same time wants us to bury long-lived waste so as not to expose it, within a few decades, to a world that will become unstable and dangerous.[6] They don't believe in the possibility of maintaining a technical *macrosystem* over a long period of time either!

Astonishing, isn't it? Nevertheless, several pathways do present themselves to the most rational among us.

Wait-and-see is the most practical approach. Acting as if you hadn't seen or heard anything requires little commitment on a daily basis. There is no need to prepare: if things turn sour, we will find people to manage the problems, that's what they are paid for. Fatalism is a somewhat extreme version of this attitude, and comprises enjoying the good times while they last, spending holidays in the Bahamas (or rather in the Maldives or Tuvalu islands, before some of them disappear) instead of wasting time in 'transition town' meetings or cheerless conferences on peak oil and the dangers of pesticides for bees. Obviously, the disadvantage of both these attitudes is the potential for a guilty conscience. Since morality is a 'value' in peril .... But of course, as you have this book in your hands and have reached the last quarter, you are not fatalistic, unless you are reading it in the Bahamas! We just need to avoid the survivalist pitfall.

This is because a bunker mentality can be very tempting: "I understand what is happening and therefore I'm going to prepare myself, my family or my local community." This is not to be confused with a healthy desire to decelerate and disengage whenever possible, by raising hens, cultivating a vegetable garden, doing without a car, choosing a less remunerative local job.... To the survivalist, the 'prepper', actions are chosen according to their financial means and inclinations: autonomous energy production and water treatment systems for some, hoarding canned food, gold coins, Kalashnikovs and ammunition for others ...

Gold is interesting in troubled times as a means – used over several millennia! – of hoarding or exchange (let's recognize that the strategy is as good as any in these times of recurrent financial crisis): a few gold coins did not hurt for oiling the wheels a little during the Second World War.

But forget having an assault rifle to protect your organic vegetable garden and to hold back the hungry hordes pouring down the roads: if a collapse of the economy occurs, to the point that the state is not able to meet the basic needs of people and they take to the roads, sophisticated industrial products will not survive either.... What are you going to do after you have used your last cartridge? Because there will always be a last cartridge, it's just a question of numbers, and by definition there are always too many invaders. Surviving in such a world means being *very* low tech, relying on an old Napoleonic musket, producing saltpetre in the cellar to make black powder, making shot from molten lead. That's very complicated. Hence some will choose a crossbow rather than a firearm. And to be safe you will need a fortified castle, not an old isolated farmhouse! There is nothing to hope for from a withdrawal into survivalism, because by contrast we will more than ever need to live together and remain a society, for better or for worse.

Hoarding canned food is also ineffective, unless you like to eat a regular diet of sardines in oil and baked beans, managing the expiry dates in the feverish expectation of a collapse that never quite arrives. Because the pace and practicalities of the inevitable changes are open to debate: will there be collapse, debacle, or perhaps adaptation in fits and starts? Maybe I am wrong, but I have the feeling that it will not be a sudden collapse, but something like a slow submergence, perhaps over the timescale of a human life as advocated by John Michael Greer.[7] Future generations may use the term 'collapse' with hindsight in the way we talk of the 'fall of the Roman empire', but historians agree that it was a political, economic and social process that lasted at least several decades, and was at times locally peaceful, and not a sudden invasion of barbarians flooding over the *limes*, the frontiers of the Roman empire.

Why? Because we ignore the very high elasticity of our consumption. In the West, we have a lot of consumption that is non-essential and that we can reduce before we need to threaten the basic needs of water, food, clothing, basic medicine, daily travel.... Even faced with a sudden or sharp increase in the price of oil, we can 'quite easily' reduce world consumption by 20–25 per cent, given the share used in private cars in Europe and the United States. Obviously, under economic pressure and in the face of sharply rising prices, people will have to carpool to get to work and to cut back on leisure activities. Of course, it will be painful (you will have to start talking to your neighbours), but it can be organized in a few months at most and without destroying the economy. What about food, with lower agricultural yields? In Europe, at least a quarter of food produced ends up in the garbage ... and by eating less meat (it will be more expensive) we release astronomical quantities of cereals! As for electricity, Japan, in its misfortune, has given us a striking demonstration of the speed at which consumption can be reduced: almost overnight, after the tsunami of March 2011 that led to the Fukushima nuclear accident, they (had to) shut down all their nuclear reactors and cut production by at least 20 per cent. How? By shutting off air conditioning in buildings, stopping escalators, restricting operations in the most energy-intensive factories.... And, so far as I know, Japan has not collapsed, although they have now restarted some thermal power plants to compensate.

The fall of the USSR was also a situation of collapse – of industrial production, of values, of local and state administrations and of public services.... Things were finally accomplished with only limited conflict, although we shouldn't underestimate the hardship that was caused and the decline in life expectancy and birth rate. Certainly, the Russians have put up with a lot, and many people still grow vegetables in their gardens today. It also depends on the scale at which you look at the subject, and from where! Because although I have difficulty imagining hordes of *sans-culottes* armed with pitchforks marching on Versailles or even the Elysée Palace, it is clear that hunger riots and revolutions could break out elsewhere, propelling ever more desperate migrants towards ever more controlled borders.

Not to mention that far too many powerful and privileged people of all kinds in control of the system have a lot to lose, far more than the average citizen or the most exploited populations of the North and the South. The oligarchic system[8] will therefore do everything it can to sustain itself, even if it means waging more and more wars in resource-rich countries and visiting extreme destruction on the environment.[9] The emergence of techniques with ridiculously poor returns – the tar sands of Alberta, Canada, production of agrofuels in temperate zones, the use of photovoltaic panels in northern zones and increasing exploration for shale oil are the precursors, and confirmation that the system will not back down from any absurdity and abuse in its quest to survive.

Moreover, believing in a sudden collapse owing to finance or to peak oil or gas reproduces the error, the Marxist original sin, of claiming that the end of capitalism is inevitable and imminent owing to declining profit rates. Experience has shown that the system's capacities of adaptation, recovery, perversion and manipulation have allowed it to survive all tests and in fact to defeat all regimes, including Soviet Russia with state capitalism and Pinochet's fascism in Chile.

If we do not try anything innovative, I lean towards a scenario of involuntary adaptation, that will be socially painful and will have a profound impact on our societies, but which will be gradual all the same. And because I don't pretend to know what will actually happen, I am saved from having to stock canned food and assault rifles, which would be incompatible with my desire for metallic sobriety. The question remains as to whether we want to try something other than this status quo, this belt tightening – as it will be at least for the majority, since, in such a scenario, the oligarchs will manage to continue to indulge, protecting their patch in an ever more strongly policed society. If we do, we will have to address at least four major issues.

The major issue of employment

Protection of employment is the number one argument used to continue to grow, despite the limits that we face, and to block any radicalism

141

in regulatory developments. It is in the name of employment that we celebrate shale gas, that we keep using plastic packaging (a sector of excellence for some parts of Europe), and that our heads of state meet dictators to sell them military equipment or civil engineering projects. And, against a backdrop of 40 years of recurrent unemployment, fear of destroying jobs is legitimate, to the point that trade unions have been known to defend seriously polluting chemical and metallurgical plants or the manufacture of weapons.

Of course, a low-technology programme would directly, but also indirectly, undermine many economic activities (through a domino effect, although it is difficult to know how it might pan out). Companies would need to close or relocate. But isn't this something they have been doing for decades, under the double blow of the negotiated reduction in customs barriers and tariffs and the drastic reduction in transport costs? Yes, but it's all a question of the rate and timing.... Relocations have meant gradual, regular but creeping deindustrialization that has allowed society to adapt in a piecemeal fashion without leaving too many people behind at the same time on the road to 'progress', with a 'sustainable' unemployment or partial employment rate (at least as seen from our parliaments) of 10–15 per cent of the active population.

We can well imagine that our aluminium production plants, which consume a lot of electricity to process imported bauxite, would soon move elsewhere if we shut down our nuclear power plants without developing an equivalent gas-based supply (fairly easy, provided we get along well with Russia or Qatar) or wind power (much more difficult if the plant cannot accommodate irregular production).

Well, OK, so what? *Que se vayan todos*, let them all go! Why, in a world of low technology and low consumption, would we need the massive production of aluminium, currently used to make car doors, window frames for skyscrapers and individual packaging for industrial palm oil cakes? We will wrap our 100 per cent organic sandwiches in a tea towel or pack them in a reusable box instead of aluminium foil, and we will buy yoghurt at the local dairy, using our small standard design glass jars that are reusable everywhere – there you go!

We can make this reflection provided that we ensure that we are able to take responsibility for the inevitable social consequences of the transition. Because, while we might not hesitate in seeing this or that harmful activity depart or disappear, there should be no question of abandoning the employees or the jobs that depend on them without offering an immediate and tangible solution, and not just the distant prospect of a more harmonious society in which times will be better. The question of employment is therefore a major one.

Homage to George Orwell

The radical change in the economic model that is essential to avoid environmental collapse is also highly desirable from a social point of view. 'Growth is employment' has been hammered into us for decades by so-called experts, and by all the media and the powerful in our societies. Orwell could have added this phrase to his famous oxymorons 'war is peace' and 'freedom is slavery'.[10]

Because this growth, which means giant retailers and agri-food, and their corollary intensive agriculture, systematic mechanization, productivity gains through increased consumption of energy and resources, and maximum concentration of activities to take advantage of economies of scale, naturally destroys many more jobs than it creates. The 'industrialized' countries have only maintained a more or less reasonable level of employment through two subterfuges.

The first subterfuge is increasing consumption. For material objects it is in the direction of everything being disposable, which is the ultimate target for consumer goods, or through the expenditures of the military-industrial complex for states (blowing up a high-tech missile is the ultimate in programmed obsolescence, the dream). For services, it is in the emergence of new needs, such as telecoms, or the financialization of what was not previously the case, such as private tutoring, which although not new has expanded enormously in recent years. The second subterfuge is to make the system more complex, in both public and private spheres, which systematically creates jobs (as managers, accountants, consultants, lawyers and so on). John Kenneth Galbraith[11] has clearly shown how any industrial group is destined to grow, internationalize and become more complex, if only to offer a *cursus honorum* – a successful career – to its senior executives.

At the same time, all out, dog-eat-dog competition prevents us from distributing work fairly, and overconsumption from enjoying productivity gains in the form of leisure, while competition between countries – in the form of fiscal, social and environmental dumping – makes any Keynesian recovery strategy impractical, because any increase in purchasing power leads to extraterritorial consumption without local benefits.

An exploration of Robinsonade?

Since factory closures have become the terror of modern times, especially in an election period, let us tackle the issue head-on and try to understand what would happen by drawing inspiration from economists' 'Robinsonades', as Marx called them. They tried – and are still trying, with complicated mathematical formulas to mask the poverty of their reasoning – to explain the behaviour and choices of 'economic actors' by reducing their number and that of their possible interactions, like Robinson Crusoe and Man Friday stuck on their island. Let us take as an illustration the disappearance of a European aluminium can factory. Of course, we could choose any other object or service that you like to consider useless or harmful: artificial fertilizers, pesticides, weapons, cars – whether 4×4 or not – disposable products, packaging, marketing or advertising services ...

Let us imagine, let us suppose that, overnight, we no longer need these cans, because we no longer have a use for them, or because we have decided, in a great eco-liberticidal impulse, to do without them because their manufacture or use is too polluting. Following a decree banning cans throughout the country (crikey!), whether made locally or imported, the factory is suddenly closing its doors. If we assume first of all that the products of the plant's suppliers, such as aluminium coils, can be stored as needed, the closure does not fundamentally change anything, within the boundaries of 'the country's society'. Besides the closed plant, the same quantities of all other products – food, clothing, housing space, transport services – will continue to be needed and produced. The number of inhabitants has also not changed, so we could maintain the same consumption, with the exception, of course, of cans that are no longer needed.

The only problem – the nub of the issue – is the unemployment of the factory staff. By losing their purchasing power, their ability to consume the production of the rest of society (food, clothing, housing ...) will be greatly reduced, the distribution of wealth will be modified. But let us imagine that these new unemployed are divided into two groups, some of them digging holes in front of the factory, others filling them up, and that they receive their previous salary for this: don't laugh, this method really was applied in France, after the February 1848 revolution, in the public works programme of the *Ateliers Nationaux*, on the basis that it was better to distribute a little money than to deal with new rebellions (and, to some extent, this method has also been applied by the World Bank in post-war circumstances, by distributing small salaries for the re-laying of laterite roads with picks and shovels). It was also proposed by Keynes, a little tongue in cheek, for the fight against unemployment. Or, if you have an entertainer's soul, we could let everyone become jugglers, dancing bear trainers or musicians, to brighten the streets of the now misnamed Canning Town.

Then nothing more would happen than the appearance of a joyful and musical atmosphere in squares and cafés where craft drinks produced locally in their free time by the former workers would be consumed, and of course bottled in returnable glass.

The conclusion of our Robinsonade exploration is that of course the decrease in consumption, and therefore in production, would endanger jobs. But all that is necessary is to find a mechanism to avoid or largely limit the social consequences of the massive closure of certain activities – by maintaining a 'lifetime' wage through extended unemployment insurance, by introducing a universal basic income, by reducing and sharing working hours, or all three at once or anything else. Those who continue to work would not be stifled by taxes, there would still be plenty of 'everything else'. It is easier said than done, of course, but the general idea would therefore be to reduce consumption of the superfluous and the corresponding jobs, while maintaining social cohesion.

Critical point: the balance of payments

Our Robinsonade analysis, however, ignores a small detail: if some of the cans were exported, no matter how useless they may be, the disappearance of the factory would change the country's consumption capacity, because they were used to pay for imported goods, to contribute to the balance of payments. This has influenced all geopolitical relations to date. In Europe, since the changeover to the euro, we tend to forget its importance, with Germany exporting for other countries. And as long as China converts its positive balance into US or European treasury bills, that is, electronic funny money, everything is fine.

We have seen how the Romans and their ostentatious spending needed the mines of Spain or Illyria to balance their trade deficit with Arabia and the East. In the 19th century, one of the most striking examples was the tea/silver/opium triptych of Sino-British trade:[12] the tea consumption of the British exploded in the last quarter of the 18th century, and the Chinese only accepted copper, gold and silver in exchange currency, so the East India Company had to find a commodity to exchange with them, in order to balance the flow of silver-metal – Europe had dried up. However, China was autonomous in terms of food, textiles and everyday products…. It was therefore the illegal opium trade – robustly supported by gunboats – that solved the problem.

Today, nothing has really changed, although international trade and more complex financial products have made the matter less clear. To import all our products, we sell cars and aircraft, cereals and services. Do you want to consume liquefied natural gas and buy new mobile phones? To balance trade, it is necessary to sell a few tanks, missiles, real estate or sports clubs to Qataris, and to welcome Chinese tourists to London or Paris and then sell them, temporarily, some expertise in gas turbines or alternators for coal-fired power plants.

It is possible that we could decrease both, and therefore why not do this in a coordinated way, in terms of consumption and required labour, but on condition that the balance between imports and exports is not too severely degraded. Ideally, we might start by giving up our many imported products, even if it means relocating production for the most essential ones.

I consume, so I create jobs ...

It is also understandable that we collectively panic at the prospect of reducing consumption, because in the miscellany of preconceived ideas, we rank highly the notion that consumption creates employment, and therefore that current unemployment is partly due to a lack of 'consumer confidence™'. Of course, this is true in a way, but we might also consider, looking at the other end of the spectrum, that it is above all a question of dividing up production quantities and working time.

Thus, instead of saying that a particular category (the elderly, the rich, the double-income DINKYs ...) with an average purchasing power higher than the population as a whole 'creates wealth' and 'creates jobs' by its consumption, we could just as well say that they create an additional workload for the active population. Without the ostentatious consumption of the rich – replace the can factory of the previous example with the whole industrial system necessary for the manufacture and maintenance of luxury yachts – we could all work a little less, provided, once again, that we know how to divide the time freed up in an equitable way. Luxury products consume scarce resources, but also and above all the time of the working and middle classes.

Economists could do the maths: how many days a year do we work to 'pay' for the private planes and leisure activities of our parasitic billionaires (sorry, our job-creating entrepreneurs)? I would say that we can make an order of magnitude estimate based on rates of return on capital. If, say, it is around 10–15 per cent, then we each put in, on average, 20 or 25 days a year of toil for the oligarchy, 'because they're worth it'. In some emerging economies, exploited through the derisory purchase price of raw materials, it is probably much more.

What can we conclude?

The return to a more leisurely consumption, based on repairable objects and short, local supply chains, could very well create many jobs. It is likely that the desirable evolution in agriculture (to smaller, more labour-intensive farms), in crafts (to the manufacture and maintenance of more durable goods), in services (to the return of the human instead of the machine, and of local trade) would provide jobs. On the other hand, reducing consumption could also destroy many of them.

What might be the overall assessment of our programme? Will we have to work more or less in a world of low technology? It would be completely misleading to try to make econometric calculations to determine the number of jobs or weekly working hours in this desirable world of windmills, bicycles and 'homemade' toothpastes. First of all, because the systemic, intertwined and interconnected nature of our economy would make the exercise futile. But above all because a calculation in terms of 'jobs' would not mean much, given several variable factors: the ratio between salaried and non-salaried jobs; the ratio between economic work and domestic work (less paid employment but more time spent at home preparing toothpaste and taking care of the vegetable garden ...); the difficulty in distributing weekly working hours; and the possibility of sharing time between paid employment and voluntary work that contributes to the good of society ...

Notwithstanding this methodological note of caution, let's still try a back-of-the-envelope calculation: because in a world that can be wonderful, but in which it will be necessary from time to time to put on a sweater or to be cold, to ride around on bicycles even if it is raining or windy, to dig up the garden or take care of the pig and hens, to replace *technoparades* by lyre and Pan flute recitals, can we at least take it a little easier than we do today?

At best, we might hope to reach a lower limit of 'two hours a day' (the title of an interesting French book of the 1970s),[13] or at least the four or five hours a day of the hunter-gatherers.[14] Unfortunately, it will probably be difficult to reach such a level because the population is now too concentrated: we have to work to produce the services that compensate for the fact that we live on top of each other, tens to thousands of inhabitants per km^2 against two or three among hunter-gatherers. We therefore need more intensive agriculture per hectare, better medicine to combat microbial proximity, a faultless drinking water and sanitation network, a transport system to carry food and to remove some of the waste.

In a world of low technology, the 'negative' effects on employment would be mainly a result of the overall reduction in needs, and the disappearance of some consumption (such as the abandonment of the majority of individual vehicles, or the reduction in the consumption of materials and energy ...). By contrast, the 'positive' effects would be from a certain de-mechanization of agriculture, industry and services; the development, appearance or reappearance of maintenance, repair and assistance services; a better sharing of working time offering the possibility of slowing down and finding a better balance between professional life, commuting time and family life; and the possibility of developing other activities, paid or unpaid, of a cultural or community nature ...

Let us as an example try to quantify these positive and negative effects on activity in different sectors of the French economy (see Figure 4.1). Agriculture has 650,000 salaried and non-salaried jobs, to which should be added those in the agri-food sector that are closely linked to the functioning of the current agricultural system, to give a total of 1.2 million. The massive development of organic farming, which is more

Figure 4.1: Work for two hours per day?

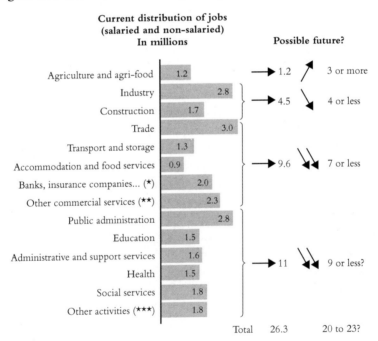

Notes:
* + accounting, legal, engineering, research ... (INSEE codes KZ, LZ, MA).
** Mainly crafts, small businesses (non-market tertiary) + INSEE codes JA, JB, JC, MB, MC.
*** Arts and entertainment, services to individuals ... (INSEE codes RZ, SZ, TZ, TZ, UZ) + tertiary non-market.
Source: INSEE Institut National de la Statistique et des Études Économiques [French National Institute of Statistics and Economic Studies]

labour intensive, should be a very important source of employment, coming on top of the partial de-mechanization of market gardening and livestock farming (but not for arable crops, probably the last thing in the world to de-mechanize). Growing on small plots of land in intensive organic market gardening creates several jobs per hectare! On the other hand, jobs should disappear in the agri-food sector, surely largely offset by the development of short distribution chains.

In total, I would say that we could see at least a tripling of the number of jobs in the sector, to reach more than 3 million, which is not particularly revolutionary when we look at a few neighbouring countries such as Italy or Poland, or indeed our own recent past. A 'return to the land' to some degree would be especially welcome to stop urban sprawl and congestion in major cities, and is therefore consistent with the rest of our policies.

It is in industry and construction that we can most easily imagine the 'negative' effects of a drastic reduction in consumption or public works and improvement programmes. However, it should be noted that the total number of jobs concerned is 'only' 4.5 million, barely 17 per cent of the current total.

There are two opposing effects: on the one hand, the outright cessation or reduction of activities, and on the other hand, the development of repair activities and artisanal activities in construction, and why not a 'partial re-industrialization' with the relocation of activities, an increase in labour intensity in certain industries using smaller units, closer to the places of consumption, although a little less productive. The result is a major unknown, as the number of jobs remaining or created will depend on the productivity rate, which is linked to energy or material consumption: 20 road-menders can be used to dig a hole or trench, or a single diesel backhoe could be used.

The number of jobs in industry will therefore depend on what we continue to consume, but above all on how we produce it, on the balance between the need to save energy and the need to avoid a return to work that is too arduous. Here again, we will have to choose our battles. Overall, the number of jobs could very well remain fairly stable, or perhaps slightly lower, compared to today.

In retailing, some activities will be heavily impacted by the decline in consumption, the relocation of activities and the development of shorter supply chains. This will be the case in supermarkets, transportation and warehousing, which alone represent 1.3 million jobs, despite the formidable efficiency of semi-trailers! This says a lot about our pressing need to transport everything, all the time, over long distances, and about the structure of the economy with cascading subcontracting that generates the need for warehouses, logistics centres, and multimodal platforms.

One may hope that professions such as those of finance, banking, accounting, legal, engineering, consulting and supervisory services will follow the general downward trend, maybe even at an accelerated rate. Starting from an enlarged tertiary system, where many professions have emerged to manage complexity but without a truly tangible final product seen by the end consumer (one can easily see that the plumber and the tax lawyer do not play in the same league), a simplification would not hurt. With the decline in consumption, many associated professions could follow: marketing and market research, lawyers and jurists specializing in competition law, accountants specializing in international mergers and acquisitions ...

Conversely, of the rest, large-scale retailing cannot survive – owing to issues linked to the use of the car, the consumption of land, the need to reduce packaging – and this could mean that many jobs reappear in local shops, as well as in certain reception and service professions, thanks to an intelligent 'de-mechanization'.

The overall balance could therefore be relative stability in industry, and a massive transfer from services to agriculture. Your customer adviser would then start harvesting something other than your bank account!

There remain the activities of administrators, education, health care, the social services, non-commercial services – a huge share of jobs today, at least a good third. How might these evolve? It will be a leap into the unknown. Beyond the age-old question of the effectiveness of public administration, how many police officers or judges will be necessary or desirable? Can their number be reduced if society has become less unequal and less ostentatious? Will we get sick less when the environment is less toxic, when most cars and junk food have disappeared? Will the slowdown reduce the number of unhappy and idle people and relieve the pressure on social services? Approximately 300,000 people currently work in the arts, entertainment and recreational professions. Why could we not double or triple this figure, by increasing the number of concerts, shows, plays and festivals?

Finally, there is the issue of the salaried or non-salaried part of our domestic activities. Several hundred thousand people work for individual employers (household staff, gardeners, and so on), and this proportion may increase or decrease depending on the type of society that we want. Will it be one, like today, where there is a large inequality in the hours people work, or another, in which we would find a way to better distribute the painful and indispensable tasks? This question could even have an impact on employment in business services: imagine if it was decided that each employee would have to take turns cleaning the toilets (oh, sweet memories of the era of conscription), even the general manager, in a collective and united effort to distribute this thankless task fairly!

In the end, our 26 million full-time jobs in France (and the 3 million unemployed, plus unregistered and part-time workers) could become, say, 20 to 23 million 'job equivalents', or, by distributing them fairly, around 3 working days per week each. There is nothing catastrophic or unimaginable here, and a perspective that could compensate for some of the efforts we make on our daily consumption. Free of cars, imported trinkets, and unwanted flat screens, we could work on our gardens, read, spend time together, restore our landscapes and cities, move more slowly – in short, we could embellish our lives.

The question of scale: lessons from British abolitionism

The second argument against such a radical change of course is that the economy is now globalized. On the one hand, many countries have made commitments (both European and global through, for example, the World Trade Organization) that we would have a hard time maintaining, leading to sanctions, reprisals or harmful exclusions; on the other hand, international trade in goods and services and capital is so intertwined that it would be impossible to implement any major programme. How could we embark on such an extravagant adventure of slowing down and relocating, while the whole world is accelerating and trade continues to increase? It is hard to imagine. And at what scale could our transition be implemented?

Might it be done on a global scale via some rescuing global governance? No matter how much we read *Libération*, the *Guardian* or the *New York Times*, how hard we try to get rid of the injustices of the world through social networks or how hard we click on 'e-petitions' to defend lost causes, basically we know that there is nothing to expect from any world organization. The few attempts, such as the Kyoto and Paris agreements on climate, while uplifting, are somewhat comical, and we will quickly come to the conclusion that we will *never* agree with 180 other countries, including some that are not very cooperative and still very influential, irrespective of the level of degradation of the planet. And while we are burying our hopes, the same is unfortunately true for the European Union in its current governance model. There are too many differences. It is not a question of trying to make an ill-advised retreat to nationalism, but of recognizing the limits of what is feasible, and, instead of trying to seek a soft consensus, of finding what we can intelligently build with our neighbours and partners.

On an individual or family level, with deliberate simplicity, we can choose moderation – why not? It has the merit that it sets an example – but how many abstainers are ready to wage an everyday struggle while embedded in a world of permanent advertising and, often, of compulsory consumption? Or is there the possibility of influencing 'markets', and production methods as an 'ethical consumer'? Let's move on from this awful buzzword. Of course, activist campaigns targeted at a product or a range of products, or even a specific brand, can pay off. But from there to transform the whole system …? Because we live in a market of supply, not demand, the economic and industrial system is driven by producers, and consumers put up with it. In contrast to the old saying, the customer is not king but buys what he finds. I searched for a long time to find butter that is both organic *and* semi-salted until it finally made it to the shelves one sunny day.

At the scale of a locality or small region, can we work on its *resilience*, its resistance to future shocks, by building on local and collective experiences and the concerns of elected municipal officials, as in the 'transition town' movement developed by Rob Hopkins and colleagues at Kinsale in Ireland and Totnes in England and which is developing internationally? It is at this scale that those who see an irreversible collusion between (evil) capitalism and the (repressive and/or hopelessly growthist) state are confined. The approach is of course interesting, it allows for awareness, exchanges, experiments and inventions, which can be useful and deployed elsewhere one day. But the means of action are necessarily limited, without access to the decisions of the public authorities.

An intermediate scale remains, which could range from a region or group of regions to a state or a small group of states whose interests and history are sufficiently convergent. In Europe, this could result, for example, in a region including France, Belgium, Germany and Italy, or the Alpine regions between France, Italy and Switzerland and so on.

At this level, it is first of all possible to discuss and make decisions. Since each country and region has its own history and circumstances, there can be no ready-made global solution. In the same way that the right *mix* of renewable energies varies according to location, depending on the available sunshine, wind, hydroelectric capacity, biomass and so on, so too can the choice of the right solutions, trade-offs and costs only be anchored in local cultures, existing urban planning and ongoing economic activities. In some cases, it will be complex and it can only be up to the local people to decide. What will be done in huge industrial areas such as the ports of Rotterdam or Shanghai, or enormous North American or Chinese conurbations? Certainly not the same as in Plymouth or Valencia. Let us find a little humility and, before giving lessons to the rest of the planet – an unfortunate paternalistic habit that we adopted with our colonies and then with 'development aid' – we should all should start by putting our own house in order.

But more than anything, what an opportunity to change things at this scale! In some ways, there was a lot to be said for good old-fashioned Colbertism.[15] The power of action of the state and of local authorities remains considerable, despite the difficulties of recent decades, in terms both of normative, regulatory and legislative possibilities, as well as in terms of the enormous volume of their purchases and the civic engagement (often) of their representatives. To take just two examples, imagine what the economic impact could be of an immediate roll-out of organic meals in canteens or public services, or of stronger mandatory requirements in the technical specifications of public buildings ...

Are you still sceptical about the possibility of a country embarking on the adventure on its own? There are some examples in history, such as the French Revolution or the abolition of slavery by the British. The latter case is exemplary. First, the United Kingdom unilaterally voted to abolish the trading of slaves in the British Empire in 1807, and their ownership in 1833. At the time, there was no shortage of economists and 'decision-makers' ready to explain that this would strain the budget of plantations and undermine the competitiveness of the country and of many small family businesses. In the decades that followed, the UK consistently applied proactive diplomacy based on bilateralism, after having renounced multilateralism as ineffective, putting pressure on other countries to submit and to vote for abolition as well. Finally, they engaged a naval force in the Atlantic, the British West Africa Squadron, to chase and challenge slave ships, increasing the difficulty and therefore the cost of the triangular trade. Of course, the British Empire was the superpower of the time, and its frenetic bilateralism was undoubtedly driven by certain hegemonic ulterior motives ...

So, what could a country, or a small group of countries, decide unilaterally? Many goods and services are traded with the world as a whole, but large parts of our economies remain domestic, and most trade is with neighbouring countries. Fundamentally, what would prevent us now from undertaking a root and branch revision of urban planning rules, to put an end to urban sprawl and reduce the need for travel? Or radically transforming our waste management system? Or launching a massive programme of support for and conversion to organic farming, to agricultural consolidation towards small plots and hedges, and to a rebalancing of arable and livestock activities? What about abandoning the senseless programmes of high-speed rail lines, highways, roads, airports and giant tunnels? Or reviewing our systems of taxation to achieve a better balance between labour, energy and raw materials? Or developing our scale of values towards a better social recognition of crafts and manual trades, or drastically redirecting public research and teaching? Objectively, nothing prevents us from doing so.

THE AGE OF LOW TECH

The equivalent of wild bilateralism could take the form of customs barriers to entry, in the form of taxes and tariffs or normative and regulatory pressures. This is another taboo to be broken: propaganda has made us equate and confuse protectionism and nationalism. In reality, protectionism is practised shamelessly by just about everyone, starting with the United States and China. Customs barriers do not mean a closed economy. Imports and exports will continue to exist, but in a form that should be similar to the old 'long-distance trading', where, in general, more raw materials were traded than finished products.

Customs barriers, to be defined and adapted, would protect local production (I can't see how small-scale livestock farming could resist the invasion of frozen Brazilian or Vietnamese chickens and so on, even with some increase in transport costs) but would also have the effect of forcing countries that want to export to align, in one way or another, to access their clients' market. And the customs officers, who have become for instance mere anti-counterfeiting agents to protect the income of those billionaires producing 'made in France' luxury goods,[16] would find a little more to occupy them. At the same time, we could easily completely revise the 'international development' aid mechanism, which continues to be based, despite a slight green varnish, on the exploitation of local resources, such as opening roads to exploit mines or forests, building dams to supply factories, and scattering a few primary school buildings everywhere.

As for the equivalent of the British West Africa Squadron, let your romantic imagination take over! The navies of countries 'in transition' could chase boats that practise intensive or deep-sea fishing, their paratroopers could carry out environmental inspections at foreign suppliers, their (few) aircraft carriers could cruise around some island tax havens (not unpleasant ideas, although they might attract some hostility). In a more measured way, we could for example imagine pressure on transnational companies that want to operate in those countries.

We live a kind of prisoner's dilemma at all levels, personal, organizational, national.... We know that we should move, but the first to move is the loser. What baker would dare to stop using paper and plastic bags if other bakeries do not? Which commercial company would dare to stop its advertising spending first, despite the fact that if they all stopped at the same time, in a sector where customer purchases are necessary anyway (such as washing powder), the consumption and market shares of each competitor would not change fundamentally? The same is true for countries, with the difference that they can use a significant regulatory arsenal to protect themselves.

Who knows, finally, if a few pioneering countries, leading the way, would not make it possible for others to follow? In the face of the impossibility of global governance, why not focus on setting an example, on trying to achieve a ripple or domino effect? It is undeniable that the activism of abolitionist societies, together with the UK's unilateral stance, soon followed by others, accelerated the end of slavery. So, let's start by sorting out our affairs, experimenting with possible ways forward, before complaining about the number of Chinese people and their coal consumption and trying to convince them not to 'develop' too much.

Cultural and moral questions

The value system: the return of manual trades

To work, the transition will have to be accompanied by a strong revaluation of manual work: marketing managers, television presenters, chartered accountants and lawyers are all more socially recognized today, in all respects (remuneration and prestige), than the shoemaker, garbage collector or stone mason. The financial trader reassures himself about his societal utility – it is always morally difficult to assume oneself to be totally parasitic – by explaining that he allows capital to be used in an optimal way. However, in a purely utilitarian vision, manual trades produce more of material value for society, more that is concrete, and which brings us comfort.

We now need to redevelop and rebalance our value system, although how remains to be seen. It will probably require that pay differentials are at least reduced. Aircraft pilots are well paid because they have the responsibility for many lives in their hands. But this is also the case for the plumber who works on the gas supply of an apartment building. With the slightest mistake, many could be killed!

The heroes of tomorrow will be farmers, rag-pickers, shoemakers, mechanics, carpenters, household appliance or computer repairers ... while bankers, accountants, lawyers, advertisers or 'market experts' will sooner or later have to disappear, or at least be significantly reduced in number. Skilled craftsmen and workers, possessors of important knowledge, were for centuries respected professions, some of them very privileged, before the ravages of Taylorism.[17]

A massive return of crafts and small industry, and therefore manual trades, would not prevent the possibility of pursuing advanced and interesting studies in parallel. Could we not imagine a society where students are not oriented towards manual occupations on the basis of poor academic performance? Matthew Crawford[18] shows well how some supposedly 'intellectual' jobs have actually become rather meaningless, a kind of white-collar proletariat, while real manual jobs require intelligence – the cabinet- or violin-maker who chooses the tree from which to make his work, the mechanic who analyses an engine failure, the builder who adapts to the unexpected – not to mention the common sense of a farmer.

One could thus graduate through study of ancient languages, then become a potter and continue to read Sophocles and Euripides in the original texts during one's leisure hours (which would potentially be a bit more numerous than today). Or perhaps we might have a society in which everyone divides their time between manual and intellectual activities: plumber-philosophers, farmer-teachers ... in a modernized version of the 18th-century English farmers who were also weavers, before they were crammed into the mills.

The hyper-specialization of work has been based on the never-ending search for productivity. However, this productivity has become destructive, requiring an increased use of energy and metals (robotization, mechanization, professional movements of specialists and so on), accelerating the need for technological equipment (computerization), excluding a large part of the population from the job market, and forcing unbridled and meaningless consumption.

On the content and objectives of education

Two fundamental failings, two fundamental movements that have transformed education for several decades, as they have transformed society as a whole, must be fought and reversed.

The first of these is utilitarianism. We no longer seek to mould and develop our citizens – if we have ever done so. In his time, the sociologist Bourdieu showed the role of education in the reproduction of elites and its effectiveness in ensuring that everyone learns their place.[19] The aim of education is to offer students the best 'employability' prospects after graduation; we educate future employees in flexibility, adaptability, how to operate in 'project-mode', and the basics of 'globish' (global English) which are now necessary for companies in a constantly changing world.[20]

Using this logic, the focus is on the subjects deemed to be the most 'useful', with a clear prevalence of scientific teaching. Meanwhile, under the pretext of efficiency, new technologies are invading schools: computers then tablets and digital whiteboards, waiting for the virtual teachers who have appeared in companies at the forefront of e-learning. There is no doubt that the steamroller is on the move, with some unbelievable future developments, such as the abandonment of the teaching of cursive writing. Well, when you can type on a keyboard.... It's already been done, or is on the way, in some states of the United States, which is as always 'leading the way'. When will we stop using that country as an example in the rest of the world?[21]

What is funny is that the children of the 1968 protest movements in Europe have often had brilliant careers in companies or government without going through this utilitarianism.... Try getting a job today as an ex-Trotskyite with sociology or philosophy studies on your curriculum vitae!

Instead of building wise heads (but with the risk of raising their critical thinking level), we make heads that are stuffed full of a wide range of 'techniques'. 'The transformation of learning into education paralyzes man's poetic ability, his power to endow the world with his personal meaning.'[22] Would it not be better to focus on fundamentals, on behavioural skills rather than know-how, on the awakening of curiosity, morality and philosophy, poetry and theatre, music, history, the arts – in short, the humanities, culture in the broadest sense, but without neglecting fundamental techniques, and especially their history? Learning ancient languages to understand our origins? Introduce traditional knowledge, such as weaving, sewing? Keep children grounded by explaining to them how the daily world works, by organizing visits to factories, workshops, landfill sites, water treatment plants, and by linking these outings to science and technology learning? Is learning to click a mouse necessary in kindergarten? I didn't learn how to at that age, and yet I'm doing pretty well.

And, finally, consumerism. The school is no longer an institution that implies rights but also duties, but a place of 'knowledge transfer'. Parent–teacher and student–teacher relations have continued to deteriorate. Teachers are no longer respected or recognized, and no doubt not helped by the arrival of a generation of teachers who have a less evident vocation, are poorly paid and often less well trained, abandoned in the face of difficulties and social violence. And while students are, more than ever, supposed to be prepared for the logic of liberal competition, effort and merit are less and less recognized while the media praise easy success.

How to make the transition desirable: long live low tech!

In a democratic society, and let us hope that it remains so, how can we convince as many people as possible of the need for and interest in change?

157

Enlightened catastrophism ...

Of course, there are lots of reports and thick books. Can the repeated announcement of future disasters convince the 'masses'? The temperatures that we will reach in a century: unfortunately, almost no one cares! The proof is that the world continues to go on as it is despite the enormity of the announcements.

I do not believe in intergenerational solidarity: far too many people on this planet are struggling to make ends meet. The number of constraints on expenditure is increasing, and things are likely to become increasingly difficult, while material growth is no longer possible, given the limited availability of energy, raw materials (for industrial production) and available land (for food production). Masked by the deflation of Chinese imports, inflation on basic products or commodities is probably back. And this time it will not be matched by wage inflation, as it was in the 1970s.

Under these conditions, let us not rely on a spirit of sacrifice. If we have to convince our fellow citizens to further tighten their belts for one, two or more generations, so that our descendants will live a little less badly, but perhaps only a little, it may be better to forget it right away and fly to the Bahamas or the Maldives.

... or the magic of fireflies!

The only solution is to make the transition desirable, to convince ourselves that change can liberate us, make us happier, make us live in a more just world, right from the start, because, as Lao Tzu told us: 'The goal is not only the goal, but the path that leads to it.'

Let us take awful cars as an example. It is not a question of riding your bike to save future generations from famine or to prevent our descendants from turning into scavengers of scrap metal, wandering hordes drawing on the huge metal stock of our deserted cities and abandoned industrial areas. If I ride my bike by myself, I am, as in the prisoner's dilemma, the loser, and I find myself with all the disadvantages – exhaust fumes, lost time, dangerous traffic – and limited advantages that only partially compensate (perhaps some moral satisfaction in 'doing the right thing'). But if everyone stops at the same time, then that's something else! The end of noise, pollution and a lot of stress. You will be able to hear birds singing, and soon, who knows, combined with stopping pesticides and unwanted night lighting, fireflies and glow worms might return! When was the last time you saw fireflies?

Of course, a world of low technology will involve the loss of a certain material 'comfort'. But we also need to understand all that we have given up in exchange, all that we have already lost since our entry into the 'Anthropocene', as described by Jean-Jacques Rousseau:

> Squalid features, unhappy wretches who languish in the infectious vapour of the mines, sooty forges, hideous Cyclops, are the objects and inhabitants which the mines substitute in the bowels of the earth, for that of verdure, flowers, the azure sky, amorous shepherds and robust labourers, who live and are happy on its surface.[23]

A reduction in material consumption could quickly make it possible to rediscover the many simple, poetic, philosophical joys of a revitalized nature that has regained its rightful place, while the reduction in stress and working time would make it possible to develop many cultural or leisure activities such as shows, theatre, music, gardening or yoga …

And, last but not least, the transition could give back hope and a reason for living, especially to the youngest, to those born after Chernobyl (and the discovery of the 'fragility of power')[24] and Thatcher ("there is no alternative"), who have switched themselves off to the tide of content of the media's 'society and environment' sections. We have one distressing announcement after another, on a daily basis: on Monday, phthalates cross the placenta, on Tuesday, GMOs spread, on Wednesday, glaciers melt or waters rise, on Thursday forests retreat, on Friday, bluefin tuna disappears (on Friday it's fish on the menu). Can we measure the cognitive effect of two decades of this? I don't know if I would have studied diligently in the face of such despair, because what's the point of graduating just before the end of the world?

No alternative, really? Well, let's think, try, experiment. Too bad if it doesn't work, at least we'll have tried something! And what a joy, in the meantime, to find a new path, a different perspective from that of a system that is on its last legs.

Finishing on a positive note

Faced with overwhelming evidence, some tentative solutions are emerging. But, given the scale of the changes to be made in all areas and in such a short period of time, ideally less than a generation, discouragement lies in wait. We are like the *Titanic*, not only because the orchestra is continuing to play while the ship is sinking but also because of the inertia in the system, our difficulty or inability to change direction to avoid the obstacle, because we have started to turn the wheel too late. Can we avoid the iceberg, in a world so interwoven, so complicated, so rooted in its (bad) habits? Many problems are tending to aggravate each other and are reinforced by positive feedback loops (see Part I). The boat is taking on water from all sides, and the breaches in the hull are getting bigger. So, is it too late?

Quite the reverse. The point of a positive feedback loop is that it works very well in both directions: if the need for raw materials is reduced, for example, the decline in energy demand will be accelerated; organic agriculture to maintain soil productivity will reduce the need for chemical inputs, and consequently energy and raw materials; soil and yield restoration will reduce pressure to clear tropical forests to compensate for depleted land.

To continue using the language of transportation, our economic system should therefore resemble more an 'oversteering' car than a liner – although this is perhaps an unfortunate metaphor. If we have the collective courage to decide to turn the steering wheel towards a set of more or less coherent measures – regulatory, fiscal, customs, cultural – then everything could be done very quickly, while providing for possible skidding.

So, let's stop being fatalistic: if transition is necessary, it is certainly possible. We have ample technical, financial, social and organizational resources. Our old planet is tired, but it has seen other challenges and we may be surprised by its ability to recover as soon as we start to reverse current trends. It's only waiting for us to take the first step …

Epilogue:
A Dream If Ever There Was One

The world has gone crazy. 'They' have gone crazy, and if it continues this way, we will eventually go crazy too! This 'progress' that we fail to stop takes daft and unexpected forms. Now, in summer, on Pornichet– La Baule beach in France, a small motorized machine cleans the sand to remove pieces of shell in order to prepare the beach for tourists.... Who could have made such a decision? Elected officials afraid of a class action lawsuit by bathers whose feet have been scratched? Consultants who have benchmarked practice in Cannes or Biarritz? And did nobody point out to them that the sand is partly composed of pieces of shell?

Things have gone too far. We might as well admit it, we won't save the planet, or at least a civilized humanity on the planet, by turning off the tap when we brush our teeth and relying on our miracle-working engineers and businessmen for the rest. It is time to take our destiny back into our own hands.

What does this mean in practice?

Let us resist the sirens on all sides who promise us that we can have the best of all worlds, with no change in comfort, mobility or consumption, while not continuing to pollute – the salesmen who claim 'how green is my method!' No, we can no longer think that the circular economy, renewable energies and a few other tweaks here and there will allow us to continue to consume like pigs, to produce and throw away like slobs.

On this Earth, unfortunately, any action has an impact. As Barry Commoner wrote: 'There ain't no such thing as a free lunch.'[1] When there was no electrical grid and sperm whale oil was still used for lighting while waiting for petroleum, life was precarious for whalers, when, before the invention of the harpoon gun, it was necessary to harpoon the Leviathan by hand. And Herman Melville's Captain Ahab cried out "For God's sake, be economical with your lamps and candles! not a gallon you burn, but at least one drop of man's blood was spilled for it."[2] Even if we pretend not to see it, nothing has really changed today.... People,

161

forests, oceans, soils, rivers, both at home and on the other side of the planet, are bending under the pressure of our expensive lives.

For my part, I try to communicate this primary truth to my children, from an early age. Maybe a little too early because it gave us "With Dad we don't take the elevator, it kills orangutans." That is slightly confused with the moratorium introduced at home on (almost) all industrial cakes and their palm oil, while my attempts to save electricity were aimed more (to stay in the mammalian domain) at the polar bear, with climate change in mind. And if, in our house, Santa Claus is deaf to calls for high-tech toys and equipment, that doesn't seem to make them sad, quite the reverse.

Everywhere, whenever possible, at all geographical scales, at home, at work, with the family, for leisure, let us slow down, simplify, disconnect, reduce. Let us favour durable, low-tech objects which will better resist future system shocks – a hand-drill can largely replace a low-cost electric drill, which will not work for more than a few dozen minutes in its entire life. Let us prefer productive activities, concrete things, earth, stone, simple pleasures.

Let's be audacious, let's dare, let's invent, let's do some DIY. And even if this is not enough to save the planet, let us not hesitate to translate into everyday life some principles to choose from, from the simplest actions to the most challenging. *Primum non nocere* (first do no harm) is a Hippocratic principle that we should apply more often, especially before we want to do 'planetary medicine' through risky geo-engineering![3] Using biodegradable plastic bags to bring back your groceries is undoubtedly a little better, sure. Or paper bags, doubtless even better. But why not reuse your bags, whether paper or plastic? Or don't use a disposable bag at all? I put my bread in a canvas bag. That's what everyone did only 30 years ago, but I must now be the only one in the neighbourhood from the weird look my baker gives me. But let us not be afraid to be ridiculed, to be pioneers, to be pedagogues, to be curious, to be exemplary, to be moralizing, to (re)engage with neighbours, shopkeepers or colleagues, to question a little our comfort and our certainties.

Let's always ask ourselves: Can I do without? Can I do less? Can I make it easier? And by the way, why do I have to do that? And couldn't I do with what already exists?

Let us sometimes choose to do as little as possible. First of all, do no harm, again, do not destroy what can still be preserved. That would already be a real revolution, as we are a long way from that today. Then perhaps don't to try to 'repair' at any price, but look carefully at the rehabilitation of certain industrial sites – especially when decontamination consists of digging up the contaminated soil only to bury it elsewhere.

Let us rediscover the virtues of regulation, even prohibition, and let us not rely solely on the 'efficiency of the market™'. Let us tackle the major items of consumption where measures can be quick, effective, exemplary, visible: reduce everything to do with the car – its use, its weight, its speed; reduce consumer waste, introduce restrictions on packaging, impose a few standard bottle sizes to introduce a universal deposit and systematic reuse; generalize the organic and the local in agriculture; regulate the sale of disposable products, toys, batteries, and so on. Let us introduce customs barriers, tariffs, regulations and standards to counter the inevitable downward levelling, downward pressure on prices, social and environmental conditions and product quality.

Let us fight, as a priority, against the irreversible, in particular land development, soil contamination, urban sprawl and the latest white elephants in public works – motorways, high-speed rail lines, tunnels and canals. Let us create associations, let us sue, let us be nuisances, let us call on our elected representatives who, out of habit, take pleasure, for their personal credit or for less obvious reasons, in the destruction of our homelands.

Let us stop producing forward-looking reports on energy mix and CO_2 emissions in 2030 or 2050, the results of which we know in advance: the future will be so much better than it is today. The world is moving too fast, it is no longer imaginable to predict what local developments and consequences may be, given the complexity of our systems, their global intertwining, but also their incredible capacity for adaptation and innovation – for better or for worse. Instead, let us make immediate and courageous decisions.

Let us resist all developments that make the system more complex, replacing the human being with the machine, in public services, companies, shops, daily activities. Let us revolt against the incredible mismanagement represented by ostentatious consumption, especially that of the richest. Let us limit, tax, prohibit and seize if necessary. Unquestionably, my recommendations are a little liberticidal. But the fundamental principle, inherited from the Enlightenment of which we are still so proud, is that the freedom of some ends where that of others begins, right? But we only have one planet: and if some people want to damage it, then we will have to have a serious discussion. How in these circumstances can we accept the existence of a class of ultra-rich?

Let us wake up and shake up our timid politicians, reduced to being (poor) managers, trying to have it all ways, overwhelmed by the complexity of the world and petrified by any major change that could compromise the results of the next elections. Their intellectual indigence and lack of perspective is truly staggering. It would be better for them to understand, as soon as possible, the frustrations and despair that are and will be generated by their attempt to preserve the status quo. Sometimes history develops in a surprising way, not always in a comfortable direction, and the current warning signs do not bode well.

Finally, instead of complaining about what we will have to give up, let us dream about how we could transform our economic system, and our lives. Let us convince ourselves that we deserve a much more charming world, a much more pleasant world, a more united and joyful society, a peaceful civilization, respectful of nature and technically sustainable. And, above all, that we have the means to do so.

To do this, I could have tried to build a utopia, and led you, for example, into a Parisian dream as Sébastien Mercier did, a few centuries ago. To talk about an 'ideal' green city with grassy paving stones, flower-filled fallow lands, ivy, wisteria, passion flowers and Virginia creeper growing on the façades, vegetable and fruit production in the old parks. To describe the occupation of avenues, cleared of cars, replaced by terraces, chess players, sports, music, art, garage sales. Be enchanted by the return of nature, bats and blue-tits in the courtyards of buildings, bees and wasps in the markets, but also – you have been warned that it's not just good things – buzzing blue flies on the butcher and fishmonger's display. Or have a few words about the new economic, cultural, moral, political system, with imaginary joyful citizens of the low-tech age.

But I do not have the talent of Mercier. So:

> In the mean time, let us endeavour to render this life tolerable; or, if that be too much, let us at least dream that it is so … O my dear countrymen, whom I have so often heard groan under that load of abuses, of which we are wearied with complaining, when will our dreams be realised? Let us then sleep on, for in that must we place our felicity.[4]

Notes

Prologue

[1] Antoine de Saint-Exupéry, *Pilote de guerre*, Gallimard, 1942, p 115, published in English as *Flight to Arras*, trans. Lewis Galantière, Harcourt, 1986.

[2] Jacques Ellul, *Le Système technicien*, Calman-Lévy, 1977, published in English as *The Technological Society*, trans. Joachim Neugroschel, Continuum, 1980.

[3] Ivan Illich, *Tools for Conviviality*, Harper and Row, 1973.

[4] E.F. Schumacher, *Small Is Beautiful: A Study of Economics as if People Mattered*, Abacus, 1978.

[5] Langdon Winner, *The Whale and the Reactor: A Search for Limits in an Age of High Technology*, University of Chicago Press, 1986.

[6] John Michael Greer has written a number of relevant books including *The Ecotechnic Future*, New Society Publishers, 2009 and *The Retro Future: Looking to the Past to Reinvent the Future*, New Society Publishers, 2017, and has a regular blog at https://www.ecosophia.net/.

[7] Kris de Decker, *Low-Tech Magazine 2007–2012* and *Low-Tech Magazine 2012–2018*, LULU Press, 2019, also https://www.lowtechmagazine.com/, Low-tech Magazine and https://www.notechmagazine.com/ No Tech Magazine.

[8] "Our house is burning ...", Jacques Chirac, Earth Summit 2002, https://www.youtube.com/watch?v=WGdh9Bzw_L0

[9] A French *département* is an administrative region, broadly equivalent to a British county. There are 95 in metropolitan France.

Part I

[1] A state of unpleasant tension caused by knowledge, opinions or beliefs about the environment, about oneself or one's own behaviour that are incompatible with each another. See: https://en.wikipedia.org/wiki/Cognitive_dissonance

[2] 'The truth pleases me; truth alone is durable.' Voltaire, *La Pucelle d'Orléans [The Maid of Orleans]*, Cramer, 1762.

[3] Bertrand Méheust, *La Politique de l'oxymore [The politics of the oxymoron]*, La Découverte, 2009.

[4] Marshall Sahlins, *Stone Age Economics*, Routledge, 1972.

[5] Franz J. Broswimmer, *Ecocide: A Short History of the Mass Extinction of Species*, Pluto Press, 2002.

[6] Alain Gras, *Le Choix du feu [The choice of fire]*, Fayard, 2007.

[7] Daniel Robineau, *Une histoire de la pêche à la baleine [A history of whaling]*, Vuibert, 2007.

8   Henry Hobhouse, *Seeds of Change: Six Plants that Transformed Mankind*, Shoemaker and Hoard, 2005.

9   John Kenneth Galbraith, *Money: Whence It Came, Where It Went*, Princeton University Press, 1975.

10 The Royal Mirror Factory of Saint-Gobain, now Saint-Gobain SA, a French multinational corporation.

11 Jean-Baptiste Fressoz, *L'Apocalypse joyeuse [The happy apocalypse]*, Seuil, 2012.

12 André Guillerme, *Les Temps de l'eau [The history of water]*, Champ Vallon, 1983.

13 Sherry H. Olson, 'Commerce and conservation: the railroad experience', *Forest History Newsletter* 9(4), 1966, pp 2–15.

14 David R. Montgomery, *Dirt: The Erosion of Civilizations*, University of California Press, 2008, and Daniel Nahon, *L'Épuisement de la terre [Exhaustion of the Earth]*, Odile Jacob, 2008.

15 Jean Mazoyer and Laurence Roudart, *Histoire des agricultures du monde [History of world agriculture]*, Seuil, 1997.

16 Marc Levinson, *The Box*, Max Milo, 2011.

17 Karl Polanyi, *The Great Transformation*, Farrar and Rinehart, 1944.

18 Clifford D. Conner, *Histoire populaire des sciences [A popular history of science]*, L'Echappée, 2011.

19 Philippe Bihouix and Benoît de Guillebon, *Quel futur pour les métaux? [What future for metals?]*, EDP Sciences, 2010.

20 William Watson Goodwin, *Plutarch's Morals*, vol. 1: *A Discourse Touching the Training of Children*, Little Brown, 1874, p 6.

21 Harold Anuta, Pablo Ralon and Michael Taylor. *Renewable power generation costs in 2018*, International Renewable Energy Agency, 2019.

22 Thomas M. Schmitt, '(Why) did Desertec fail? An interim analysis of a large-scale renewable energy infrastructure project from a Social Studies of Technology perspective', *Local Environment* 23(7), 2018, pp 747–76.

23 Mark Z. Jacobson, Mark A. Delucchi, Zack A.F. Bauer, Savannah C. Goodman, William E. Chapman, Mary A. Cameron et al, '100% clean and renewable wind, water, and sunlight all-sector energy roadmaps for 139 countries of the world', *Joule* 1(1), 2017, pp 108–21.

24 Vaclav Smil, 'What I see when I see a wind turbine [Numbers Don't Lie]', *IEEE Spectrum* 53(3), 2016, p 27.

25 See: https://negawatt.org/en

26 Our World in Data, 'Global renewable energy consumption', https://ourworldindata.org/renewable-energy

27 'Leading scientists set out the resource challenge of meeting net zero emissions in the UK by 2050', press release, 5 June 2019: https://www.nhm.ac.uk/press-office/press-releases/leading-scientists-set-out-resource-challenge-of-meeting-net-zer.html

28 World Bank Group, *The Growing Role of Minerals and Metals for a Low Carbon Future*, World Bank, 2017.

29 Serge Latouche, *Bon pour la case [A piece of junk]*, Les Liens qui Libèrent, 2012.

30 Pièces et Main d'œuvre, *Aujourd'hui le nanomonde [Today the nanoworld]*, L'Echappée, 2008.

31 Edgar J. DaSilva, 'The colours of biotechnology: science, development and humankind', *Electronic Journal of Biotechnology* 7(3), 2004, pp 01–02; Mayara C.S. Barcelos, Fernanda B. Lupki, Gabriela A. Campolina, David Lee Nelson and

Gustavo Molina, 'The colors of biotechnology: general overview and developments of white, green and blue areas', *FEMS Microbiology Letters* 365(21), 2018, provides a recent discussion.

32 Indeed, European Union legislation is going this way: https://ec.europa.eu/commission/presscorner/detail/en/qanda_20_419

33 Although the RepRap project has devised printers that can make some of their own parts: https://en.wikipedia.org/wiki/RepRap_project

Part II

1 *Les Visiteurs* is a French fantasy comedy film in which a 12th-century knight and his servant travel in time to the end of the 20th century and find themselves adrift in modern society.

2 Kris de Decker even proposes using heated clothes: https://www.lowtechmagazine.com/2013/11/heat-your-clothes-not-your-house.html

3 Hermann Melville, *Moby-Dick* or *The Whale*, Harper and Brothers, 1851.

4 Ivan Illich, *Tools for Conviviality*, Harper and Row, 1973.

5 The Citroen 2CV: cleantech from the 1940s, https://www.lowtechmagazine.com/2008/06/citroen-2cv.html

6 'Scandal of Europe's 11m empty homes', *The Guardian*, 23 February 2014.

7 Adam Smith, *An Inquiry into the Nature and Causes of the Wealth of Nations*, Thos. Nelson and Peter Brown, 1827.

8 Matthew B. Crawford, *Shop Class as Soulcraft: An Inquiry into the Value of Work*, Penguin, 2009, quoting Keith Sword, *The Legend of Henry Ford*, Rinehart, 1949.

9 Émile With, *Les Métaux [Metals]*, Furne, Jouvet et Cie (no date; circa 1860), p 354.

10 Émile With, *Les Métaux*, p 355.

11 Nicolas Lambert, *Avenir Radieux, une fission française [Radiant future, a French fission]* (theatrical show).

12 Matthew B. Crawford, *Shop Class as Soulcraft*.

Part III

1 Stephen J. Scanlan, 'Feeding the planet or feeding us a line? Agribusiness, "grainwashing" and hunger in the world food system', *International Journal of Sociology of Agriculture & Food* 20(3), 2013, pp 357–82.

2 Nikolaus Geyrhalter, *Notre pain quotidien [Our daily bread]*, KMBO, 2007, or Erwin Wagenhofer, *We Feed the World*, Zootrope films, 2007.

3 A. Muller, C. Schader, N. El-Hage Scialabba et al., 'Strategies for feeding the world more sustainably with organic agriculture', *Nature Communications* 8(1), 2017, https://doi.org/10.1038/s41467-017-01410-w

4 Jacques Ellul, *The Technological Society*, Continuum, 1980.

5 Daniel Pauly, *Vanishing Fish: Shifting Baselines and the Future of Global Fisheries*, Greystone Books, 2019.

6 Bernard Charbonneau, *L'Hommauto [Car man]*, Denoël, 1967, p 123.

7 Ivan Illich, *Energy and Equity*, Harper and Row, 1973.

8 Or: 280 mile/Imperial gallon, 235 mile/US gallon.

9 Buildings designed in the early 19th century by French philosopher Charles Fourier for self-contained utopian communities to live and work together.

10 Charbonneau, *L'Hommauto [Car man]*, p 74.

11 Immanuel Kant, *Grundlegung zur Metaphysik der Sitten*, 1785, available in many English translations. *Groundwork of the Metaphysics of Morals*, 2nd edn, trans. Mary Gregor and Jens Timmermann, Cambridge University Press, 2012 used here.

12 Hervé Kempf, *Comment les riches détruisent la planète [How the rich are destroying the planet]*, Seuil, 2007.

13 David Reinsel, John Gantz and John Rydning, *Data Age 2025, The Digitization of the World*, IDC (International Data Corporation), November 2018, https://www.seagate.com/files/www-content/our-story/trends/files/idc-seagate-dataage-whitepaper.pdf

14 The Y2K bug was the problem with the formatting and storage of dates in some computer software that was expected to occur at the beginning of 2000.

15 Although the full site with edit history is considerably more: https://en.wikipedia.org/wiki/Wikipedia:Size_of_Wikipedia

16 See: https://meta.wikimedia.org/wiki/Wikimedia_servers

17 Aldous Huxley, *Brave New World*, Chatto and Windus, 1932.

18 Alfred Sauvy, 'Subsistances et matières premières' [Subsistence and raw materials], *Revue Economique* 3(1), 1952, p 91.

19 David Graeber, *Debt: The First 5,000 Years*, Melville House, 2011.

20 A tontine is an investment plan for raising capital combining features of a group annuity and a lottery. They were devised in the 17th century and are still common in France.

21 *Le Monde Diplomatique*, May 2013: review by Greg Campbell, of the film *Diamants de Sang [Blood diamonds]*.

22 Jared Diamond, *The Third Chimpanzee*, Hutchinson Radius, 1991.

23 Paul Valéry, *Regards sur le monde actuel [Views on the current world]*, Stock, 1931, p 63.

24 Michel Rocard, a former French Prime Minister. See (in French): https://fr.wikipedia.org/wiki/La_France_ne_peut_pas_accueillir_toute_la_mis%C3%A8re_du_monde...

25 Jean-Claude Michea, *L'Empire du moindre mal [The empire of the lesser evil]*, Climats, 2007, and *La Double Pensée [Double thinking]*, Flammarion, 2008.

26 Laurens van der Post, *The Lost World of the Kalahari*, Hogarth Press, 1958.

27 Dominique Le Brun et Collectif, *Les Baleiniers, témoignages 1820–1880 [The whalers, testimonials 1820–1880]*, Omnibus, 2013, p 301.

28 Richard Fleischer, *Soylent Green*, film, Metro Goldwyn Mayer, 1973.

29 Plutarque, 'Vie de Solon', in *Les Vies des hommes illustres [Lives]*, trans. Jacques Amyot.

30 Harmut Rosa, *Accélération [Acceleration]*, La Découverte, 2010.

31 Martin Monestier, *Histoire et bizarreries sociales des excréments [The history and social quirks of excrement]*, Le Cherche midi, 1997.

32 Lewis Mumford, *The Myth of the Machine*, Secker and Warburg, 1967.

33 Association négaWatt, *The 2017–2050 négaWatt Scenario Executive Summary*, January 2017, https://negawatt.org/IMG/pdf/negawatt-scenario-2017-2050_english-summary.pdf

Part IV

1 Serge Latouche, *Bon pour la case [A piece of junk]*, Les Liens qui Libèrent, 2012.

2 Joseph A. Tainter, *The Collapse of Complex Societies*, Cambridge University Press, 1988.

3 Hervé Kempf, *Comment les riches détruisent la planète [How the rich are destroying the planet]*, Seuil, 2007.
4 Gilbert Rist, *Le Développement, histoire d'une croyance occidentale*, Presses de Sciences Po, 1996, published in English as *The History of Development: From Western Origins to Global Faith*, trans. Patrick Camiller, Zed Books, 1997.
5 Harmut Rosa, *Accélération [Acceleration]*, La Découverte, 2010.
6 Michael Madsen, *Into Eternity*, Dogwoof, 2011, is a documentary film that follows the construction of the Onkalo waste repository at the Olkiluoto Nuclear Power Plant in Finland.
7 John Michael Greer, *The Long Descent*, New Society Publishers, 2008.
8 Hervé Kempf, *L'Oligarchie, ça suffit, vive la démocratie [Oligarchy, that's enough, long live democracy]*, Seuil, 2011.
9 Bertrand Méheust, *La Politique de l'oxymore [The politics of the oxymoron]*, La Découverte, 2009.
10 George Orwell, *1984*, Secker and Warburg, 1949.
11 John Kenneth Galbraith, *The New Industrial State*, Houghton Mifflin, 1967.
12 Henry Hobhouse, *Seeds of Change: Six Plants that Transformed Mankind*, Shoemaker and Hoard, 2005.
13 Collectif Adret, *Travailler 2 heures par jour [Work for 2 hours a day]*, Seuil, 1977.
14 Marshall Sahlins, *Stone Age Economics*, Routledge, 1972.
15 Jean-Baptiste Colbert was the French minister of finance under Louis XIV. Colbertism is a set of economic practices related to mercantilism.
16 François Ruffin, *Leur grande trouille [Their big funk]*, Les Liens qui Libèrent, 2011.
17 Taylorism is a scientific theory of management based on the analysis of work and workflows, named after US mechanical engineer Frederick Taylor (1856–1915).
18 Matthew B. Crawford, *Shop Class as Soulcraft: An Inquiry into the Value of Work*, Penguin, 2009.
19 Pierre Bourdieu and Jean-Claude Passeron, *La Reproduction*, Éditions de Minuit, 1970, published in English as *Reproduction in Education, Society and Culture*, Sage, 1977.
20 Luc Boltanski and Eve Chiapello, *Le nouvel esprit du capitalisme*, Gallimard, 1999, published in English as *The New Spirit of Capitalism*, trans. Gregory Elliott, Verso, 2005.
21 Emmanuel Todd, *Après l'empire*, Gallimard, 2002, published in English as *After the Empire: The Breakdown of the American Order*, Columbia University Press, 2006.
22 Ivan Illich, *Tools for Conviviality*, Harper and Row, 1973.
23 Jean-Jacques Rousseau, *Œuvres posthumes*, Genève, 1781, published in English as *The Confessions of J.J. Rousseau, Citizen of Geneva: Part the First. To which are Added, The Reveries of a Solitary Walker*, vol. 2, 3rd edn, G.G. and J. Robinson, London, 1796.
24 Alain Gras, *Fragilité de la puissance [The fragility of power]*, Fayard, 2003.

Epilogue

1 Barry Commoner, *The Closing Circle: Nature, Man and Technology*, Knopf, 1971.
2 Herman Melville, *Moby-Dick or The Whale*, Harper and Brothers, 1851, p 208.
3 Clive Hamilton, *Earthmasters: Playing God with the Climate*, Allen and Unwin, 2013.
4 Louis-Sébastien Mercier, *L'An 2440: Rêve s'il en fut jamais*, Paris, 1771, translated into English as *Memories of the Year Two Thousand Five Hundred*, trans. W. Hooper, MD, W. Wilson, 1772.

Index

Note: Page numbers in *italics* indicate figures and tables.
Page numbers followed by 'n' (e.g. 165n) refer to notes.